Datapoint

The Lost Story of the Texans Who Invented the Personal Computer Revolution

Lamont Wood

Photo credit John Dyer

Datapoint: The Lost Story of the Texans Who Invented the Personal Computer Revolution
Copyright 1-292673341 ©2010 by John Frassanito & Associates

All rights reserved. No part of this book may be reproduced or transmitted in any form or by any means without written permission from the John Frassanito & Associates

ISBN 978-0-615-52056-8

Table of Contents

Introduction		5
Chapter 1	The Debut: July 10, 1971	9
Chapter 2	The Gathering: 1968	15
Chapter 3	Tooling the Datapoint 3300: First Half of 1969	27
Chapter 4	Inventing the Personal Computer: Second Half of 1969	37
Chapter 5	Sparking the Microchip: Late 1969, Early 1970	47
Chapter 6	The 1201 Chip: Spring 1970 Through 1971	57
Chapter 7	CTC Makes the 2200: 1970-1972	69
Chapter 8	CTC's Finances: 1970-1972	83
Chapter 9	"The Worst Business Decision in History"	89
Chapter 10	Professionalization: 1972-73	95
Chapter 11	Gus Roche's Second Act, and Final Act: 1974-75	105
Chapter 12	The Microprocessor Market Blossoms: 1972-1977	115
Chapter 13	Invention of the LAN: 1975-1977 (and beyond)	121
Chapter 14	The Salad Days: 1977-1981	133
Chapter 15	The Debacle: 1982-1985	143
Chapter 16	The Raider: 1985-87	155
Chapter 17	Converging Dangers	165
Chapter 18	Death by Irony: 1987-2000	177
Chapter 19	Roll Call	187
Appendix A	The Patent Soap Opera	189
Appendix B	Timeline of Events	197
Appendix C	Datapoint Financial Results	207
Appendix D	Processor Ratings	209
Appendix E	Computer Terminal Corporation Business Plan	213
Bibliography		241

Introduction

More than a billion PCs[1] are in use today. The PC's popularity surely ranks it as among the most successful of mankind's inventions, along with the telephone, the automobile, and the printed book. Yet, as far as knowing where the PC came from, we are left with tales reminiscent of the creation myths of aborigines. Like those myths, the tales have been handed down, unquestioned, from source to source, in this case in the form of derivative, repetitive histories that extol Silicon Valley luminaries like Bill Gates and Steve Jobs. They might mention that the PC's success is partially tied to it being a grass-roots standard that is not beholden to any one computer maker, founded on a line of microprocessors called the x86 line. They might mention that all x86 machines have a common ancestor called the Intel 8008, which came along after the Intel 4004, which was the first microprocessor. Sometimes they mention that the 8008 owed something to a sort of pre-creation entity called CTC, later called Datapoint. Then, typically, they go back to extolling Gates and Jobs.

Yet, the history of the PC hails back (at this writing) slightly more than 40 years. Many of those involved in its creation remain among us and can be interviewed by anyone willing to track down their phone numbers. That is what was done for this book, with the knowledge that further delay would steadily add to the number slipping beyond reach, until only silence remained. Apart from some magazine articles, a doctoral dissertation, and some interviews produced by the Computer History Museum in Mountain View, CA, many had not been previously consulted.

The result is a story far more complex—and unlikely—than any creation myth. It turns out that the actual inventors of the PC were three hard-driving designers and engineers from the space program who founded Datapoint (originally Computer Terminal Corp., or CTC) in obscure, semi-colonial San Antonio, Texas. They developed and sold the first desktop personal business computer based on plans that originated as far back as 1968, when Bill Gates and Steve Jobs were still in middle school. They cloaked their plans for that computer in euphemisms, not to avoid tipping off competitors, but to avoid alarming their backers.

A side effect of their coyness is the myth (still circulating after four decades) that they invented the personal computer by accident, having set out to make a "programmable terminal." But the record shows that they intended from the start to make a personal, desktop computer. The question was how they were going to market this new thing, and "programmable terminal" was an obvious option.

After many setbacks (mostly financial) they succeeded in creating and marketing their computer, the Datapoint 2200. After further setbacks, its heart was rendered as the mass-produced Intel 8008 microchip, which gave rise to the x86 dynasty, which became the heart of the PC standard. Consequently, all billion-plus PCs in use today are updated versions of that first computer produced by Gus Roche, Phil Ray, and Jack Frassanito. Their names are on U.S. design patent 224,415, filed November 27, 1970, which describes the machine.

It also turns out that the 4004 (the supposed first microprocessor) had nothing to do with

---

[1] By PC, we are, of course, referring to personal computers that can run Microsoft Windows software, whose architecture can be traced back to the original IBM PC of 1981.

the 8008. In fact, the 4004 came out before the 8008 mostly by accident. The only thing they had in common was Intel. Their names (which imply an association via numeric progression) are misleading. Meanwhile, Intel was originally uninterested in getting involved with the project that became the first computer microprocessor—for what were, actually, very good reasons—but in the end profited immensely from Datapoint's invention.

Datapoint went on to join the Fortune 500 and be the biggest civilian industry employer in San Antonio, with nearly 9,000 employees (in 27 countries) in the early 1980s. Then, for reasons that can be seen as either complicated or starkly simple, it was crushed by its own invention.

Weirdly, neither Intel nor Datapoint bothered to patent the microprocessor itself—although others did, producing a legal soap opera (laid out separately in Appendix A) that left some of the participants wary of telling their stories, decades later.

There were points on which memories did not agree—and points on which they clashed outright. Where they could not be reconciled, it is so stated. What matters is that, possibly for the first time, the surviving principals behind the Datapoint 2200 and the Intel 8008 are on record.

Of course, the microcomputer would have been created eventually, with or without Roche, Ray, or Frassanito—the industry's ability to put more and more components on a chip meant that someone was eventually going to fabricate one. But without these dreamers who broke from the herd, it would have been trotted out by the existing computer vendors to serve their narrow marketing aims—it would have been just another custom chip buried in an expensive, brand-name machine. There is no guarantee that the PC, as a grass-roots standard under the control of no single vendor, that let anyone make a computer and anyone else write software for it, would have emerged. It might have been years, if not decades, before the microprocessor took on a life of its own. The modern digital environment would certainly not have emerged as it did.

But this is not a story about conflict or controversy. There was no race to invent the PC. What emerges instead is a love story, involving people who were in love with technology, and in love with making it useful. They were people who could not imagine more fun than going to work in the morning and inventing things, grudgingly going home late that night—or the next week. Today, a billion people can turn on their personal computers and experience the results of that enthusiasm.

Meanwhile, thousands of people were involved in the Datapoint saga, and are aware that history has largely overlooked both Datapoint's important contributions as well as the drama of its eventual downfall. Their memories are not always happy, and there were those who resist bringing Datapoint's story to mind. Perhaps this book can help them reestablish a positive link to the past. If nothing else, this book can show what happened to Datapoint, and why.

Finally, every writer is indebted to his sources. A partial list of those I'd like to thank include (in random order) Austin Roche, Michael Fischer, Chris Roche, Ed Gistaro, Jack Frassanito, Stan Mazor, Gerald Mazur, Vic Poor, Amy Wohl, Robert Metcalfe, Jonathan Schmidt, Harry Pyle, David Monroe, Gordon Peterson, Federico Faggin, Herb Baskin, Hal Feeney, Ted

Hoff, Gerry Cullen, Bob McDowell, Ted Nelson, Chrisa Norman Scoggins, Jeff Jackson, Bob McClure, Richard Erickson, Joel Norvell, Egil Juliussen, Chuck Miller, John Murphy, and Michael Knoop.

Chapter 1

The Debut: July 10, 1971

It was 1971, shortly after the Fourth of July holiday. The war in Vietnam was unfolding. Apollo 15 was readying for lift-off to the moon. President Nixon had just signed the 26th Amendment to the U.S. Constitution lowering the voting age from 21 to 18. The Watergate scandal was 11 months in the future. Unknown to the citizens of San Antonio, Texas, the drought they had been experiencing for the last two years was about to end with, as usual, floods.[2]

The offices of the Computer Terminal Corp. (CTC) were an air-conditioned oasis in the summer heat of the rolling hills and scrubby oak trees on the elevated outskirts of northwest San Antonio, along the southern edge of the Texas Hill Country. Outside, roadrunners darted in and out of the bushes that bordered the parking lot, and deer sometimes wandered into the open. Vantage points above the trees revealed a spreading downhill vista that stretched south nine miles to a cluster of tall buildings that marked downtown San Antonio. Well out of view from that distance, the famous Alamo and the River Walk lay tucked between those buildings.

The heavily treed land between the office and the downtown was mostly occupied by Anglo neighborhoods, whose residents spoke English at home. To the west of downtown was a flatter, less green district mostly occupied by Hispanic neighborhoods whose residents often (but not always) spoke Spanish at home. The language of the billboard advertising typically reflected the linguistic leanings of the neighborhood, but billboards along major boulevards could lean either way, so that even the most insular Anglophile eventually figured out that Budweiser was El Rey de las Cervezas. The other major districts of the city were a cultural toss-up, reflecting the presence of the military and its transient members—attracted by the generally clear weather, there were three active Air Force bases in the area, one Army base, and numerous support facilities. There was one bedroom community with, supposedly, 55 retired generals. The lawns were immaculate.

Inside the CTC building, the founders were holding court. After three years of struggle, they were ready to show off their creation to the press. During those three years they had nearly gone out of business more than once, avoiding disaster most recently through an agreement to partner with a multinational corporation. They had, in fact, weathered a recession in the computer industry that had driven two of the eight mainframe vendors out of the business.

But little details like that were not allowed to be a wet blanket that day, as they hosted a visit from a reporter from the San Antonio Express-News, one of the two local daily newspapers in the city at the time. In fact, this was apparently their first serious attempt at local press relations. They showed off their creation on a table beside a printer, and later, when the picture appeared in the newspaper, the caption writer was careful to specify which was the

---

2 http://www.srh.noaa.gov/ewx/html/wxevent/Climate_Narratives/julclimate.htm, referenced April 28, 2009.

printer and which was the new system.[3]

The new system was called the Datapoint 2200. In terms of a functional description, there appeared to be some question as to what to call it, and variations of "terminal" prevailed, there and elsewhere, perhaps reflecting the corporate name. The reporter described it as a "terminal with a built-in computer."

No one called it a personal computer, the term not having been invented yet.

That is what it was, however—the first mass-produced, desktop personal computer. It had a keyboard, screen, mass storage, internal memory, a processor, an operating system, communications facilities, and a price that justified personal use. All the major features of a modern PC were there, admittedly in sometimes rudimentary form.

But there's more to it than that—in the next few years, plenty of other people came out with desktop computers. Almost all of them used architectures unique to themselves, and could not run software from any other model. They are now found mostly in museums, the public having embraced the industry-wide PC architecture that no one could claim to own. But the Datapoint 2200 was different—it still survives, because it became the basis of the subsequent PC architecture. A recognizable shadow of that machine is present in every modern PC. In fact, every modern PC can be thought of as a modernized version of the Datapoint 2200.

It all happened because, literally while they were talking that day in San Antonio, steps were under way at the facilities of another new venture, in California, called Intel, to design the processor circuitry of that machine into a single chip. That chip, the 8008, would reach the market in a little less than a year, and draw considerable attention. An upgrade that came out two years later, called the 8080, turned out to be much easier to design into a circuit board. In 1975 it began appearing in cheap systems aimed at hobbyists—in other words, as a home, or personal, computer. As further upgrades appeared it became part of the IBM PC and its clones. As the PC became an industry and then a cultural fixture—almost replacing the computer industry as it previously existed—it created its own heroes and titans And those heroes and titans were soon overcome with nearly complete amnesia about that semi-rural electronics factory in San Antonio.

Judging from the coverage they got, CTC's founders spent more time that day talking about their previous product, the Datapoint 3300. It may have been easier to talk about, and anyway it was the product they had founded their company on. It was, simply, a computer terminal, designed to offer all the features that a computer terminal ought to have—but at the time often didn't. The features included a sharp, clear display, and a keyboard that was attractive to anyone accustomed to an electric typewriter. Plus, it could fit into a quiet office environment. Most especially, you could plug it into the same data port formerly used by a junky, old, loud, unreliable, paper-eating, ribbon-breaking electromechanical Model 33 Teletype, and the computer would not know the difference.

The computer market had embraced the Datapoint 3300. Anyway, Model 33 Teletypes (mainstay of the then-booming computer time-sharing business) were in short supply.

---

3 San Antonio Express-News, July 11, 1971, "Making a Better Data Mousetrap," by Bill Barnes. There was no recorded coverage by the other local daily newspaper, the San Antonio Light.

When it began shipping the units in September 1969, CTC had orders for 1,400 Datapoint 3300s.

But, they told the reporter, it was the Datapoint 2200—or something like it—that they had wanted to build all along. The Datapoint 3300 was just a platform on which to establish their company, and a product whose purpose was easier to explain to venture capitalists—their original idea had been "too radical," one of the founders explained. Now that they had a viable company, with manufacturing and service facilities, they could go on and build what they really wanted to build.

Their idea would catch fire—in fact, their eventual impact on civilization may be as profound as that of Gutenberg. But it did not catch fire in a way that would benefit their company, and not before the founders were themselves forgotten.

However, their remarks about their original intentions barely caught the reporter's attention that day. He proved more intent on describing CTC by the numbers: it then had an annual payroll of $2.73 million, with 275 employees, 205 of whom are in manufacturing, 35 of whom were salesmen, and 35 of whom were field service technicians. They had seven field offices. Datapoint 2200s had already been installed in 40 different companies. The firm sat on a 14 acre plot, and its headquarters had 63,000 square feet of manufacturing space.

The reporter spent even less space describing CTC's founders. But, as this book will show, they were the founders of the personal computer revolution. The development of PC technology can be traced directly back to what they did, and the machine they were showing that day. Microsoft got its halting start in 1975, and Apple sold its first system (a crude circuit board) in 1976. The founders of CTC began manufacturing a fully functional, stand-alone desktop personal computer, at a price that justified use by one person, in November 1970. (They then waited eight months before calling the local newspaper, however.)

Inside and Outside

The founders, who were showing off their pride and joy that day, were Gus Roche and Phil Ray. They were engineers with long experience in the electronics industry, and had met while working on NASA projects. They were not the myopic, meek engineers of popular culture—in school, they both got in trouble for fooling around with homemade explosives. Their love of fast cars would later get them both in further trouble. Roche pursued cave diving as a hobby until CTC's key-man corporate insurance policy forbade it, and then switched to sailboat racing. Ray seemed intent on upholding the reputation of Texans everywhere.

Presumably, they were attracted to electronics because it was then a pioneering field, and offered all the excitement that goes with pioneering. After they got satiated with the space program's form of excitement, they moved on to the ultimate challenge—founding a company. Later, when that "wasn't fun anymore," they moved on, together, to the other ventures, leaving their creation (then called Datapoint Corporation) to its own adventures— and there turned out to be no shortage of those.

Along the way, they created the personal computer—and it created the modern world.

Usually, in such a partnership, one member is the "inside man" and the other is the "outside man." The outside man makes the high-concept speeches to the bankers and backers. If he is not the source of the guiding conception that drives the enterprise, he has made it his own—and then made it his mission to win over others. The inside man is the one who gets things done, and handles the grubby details of running the enterprise.

By all accounts, Roche was both the inside man and main source of ideas. The idea of the desktop personal business computer that led to the Datapoint 2200 apparently originated with him (although others designed it.) He consciously assigned Ray the outside role. Later he would make Ray the president of CTC while he, Roche, remained vice president. (Ray would always say that he and Roche drew straws to see who would be president, with the loser taking the title.) Meanwhile, a business partner would complain about Roche's Svengali-like control over Ray. It was Roche who explained to the reporter in 1971 that the idea for the 2200 had been "too radical" to found a company on.

Most importantly, unlike many engineers turned entrepreneur, Roche and Ray avoided the "black box syndrome." Many engineers, thrust into management, have concentrated on what they knew, and proved uninterested in involving their enterprise in anything outside that comfortable box. Perhaps it was from their risk-taking personalities, or because they were a partnership and could share risks. Or perhaps their attitudes reflected their experience in the space program where everything they did was new. Regardless, once they founded CTC they immediately set off into uncharted territory, and actively sought help to bolster their own skills. They assembled hand-picked technologists to make their dream come true, and the results remain with us.

In hindsight (especially for a PC user) their achievement might seem inevitable. As this book will show, it was anything but.

Origins

Jon Philip "Phil" Ray was born in 1935, but later gave differing accounts as to exactly where. Raised in Texas, sources agree only that he was a child of poverty who did not know his father. In his heyday he liked to go with friends to fancy restaurants and order whatever entrée was the most expensive, without even asking what it was. He would then eat no more than half of it. When expensive wine was brought out, he thought it was a great joke on the maitre 'd to pretend to hate it, and spit it out. He played Willie Nelson music, and he had friends from all walks of life—visitors to his house never knew whom they were going to encounter.

He kept a microscope in his office and actually studied chemistry as a hobby, at one point using a dictionary to translate a Russian chemistry book that he wanted to consult. One of the cars he drove to work was a Porsche racer that was not street-legal, and whose soft tread-less tires tended to sag in hot Texas parking lots. He may have been one of the first people to explore the use of lasers as toys (to the dismay of onlookers.) His practical jokes including installing a rheostat in his secretary's wall clock so he could adjust the time from his desk, or make it run slower as quitting time approached. He also once rigged her typewriter with string to appear to type HELP while unattended.

When CTC was founded Ray was married and had two children. He subsequently got a divorce and married a CTC office employee. She was with him when he died in 1987—of cancer. Indeed, he apparently smoked almost non-stop during his time at CTC. Former employees recalled how he would place lit cigarettes on furniture as he moved about the facilities. He placed them upright on the filter end, like little lampposts, and would leave them as he moved on to the next person. At the end of the day, a cluster of them on a person's desk was evidence of the boss's attention. (Obviously, this was long before smoking inside public facilities was effectively banned.)

He liked fancy cars and on weekends turned the parking lot of his factory in a rally raceway. He raced to win. He also entered road rally races, held on regular roads with the competition based on accurate timekeeping and navigation. With a timer for an Atlas missile connected to the speedometer, and various control boxes, he set new standards for accuracy.

It was clear that he was in it for the fun. "He was a cowboy, an unrestrained Texan," recalled Jonathan Schmidt, who was later brought in to help create the Datapoint 2200.

But when things got serious, so did he. One night a fire broke out in the Datapoint factory. Ray was living next door, and showed up at the scene in a t-shirt and hastily donned pants and managed to talk the fire captain into putting away the fire hoses—pumping water into the building would have ruined the company's inventory of electrical components and perhaps put it out of business. He went on the roof and found the hot spot, and determined where the fire was inside the building and how big it must be. He estimated that two men with oxygen masks and chemical fire extinguishers could put it out—and he would go in if they wouldn't. They did.

Ray graduated from the University of Texas with a bachelor of science in electrical engineering in 1957. A teacher remembered him as the best lettering artist he'd ever seen, producing eye-popping posters. Subsequent employers included Texas Instruments, International Data Systems, and NASA contractor General Dynamics' Dynatronics division.

His business partner, Austin Oliver "Gus" Roche III, was completely different material, but was as colorful in his own way. Six years older than Ray, he was born November 11, 1929, in Brooklyn, and raised in Indianapolis. The "Gus" nickname came from a teacher who misread his first name as Gustin because his written A looked like a G. His father was a mechanical engineer specializing in steam engineering for fixed infrastructure. As a teenager Gus learned gunsmithing and (from a race driver who lived in the neighborhood) Maserati repair. His father forbid him to take flying lessons—so he did anyway, under the guise of a weekend job, and one day his father noticed him flying overhead, upside-down, thumbing his nose.

Although he did not pursue flying as either a career or a hobby, he did end up in the U. S. Air Force from 1948 to 1952, and was on Eniwetok during the first H-bomb test in 1952. He was exposed to more radiation than was anticipated, and two people in the same building died of cancer. This incident came back to haunt Roche's family in 1975.

He subsequently graduated from Purdue University in 1956 with a bachelor in electrical engineering. Subsequent employers included Radiation Inc., Emerson Research, General

Dynamics' Dynatronics division, Martin-Marietta, and then General Dynamics' Dynatronics division again. The last three positions involved the space program.

At the time that CTC was launched, Roche was married to Barbara Sue Hanna, an artist who participated in avant-garde movements. They had four children, two girls and two boys. Visitors likened Roche's house to that of a college professor's, with bookshelves, fine wine, and John Cage music—although he worked such long hours that he was apparently not there a lot.

While he would enter the weekend car races in the parking lot, he was more interested in using that space for monthly outdoor parties for the employees, paying for beer, tamales (the local cultural equivalent of hot dogs) and a band. He did sail competitively—and sailed to win.

He believed in motivating the people around him through one-on-one interactions and other special efforts to get to know them—and probe them. Superficially, this meant eating and drinking with them, often late into the night, but there was always an agenda that was clear only to him—it was as if he were engineering the engineers as well as the products. Some saw his approach as overbearing, and it drove them away. It made acolytes of others—decades later, some former associates still commemorated his birthday, and talked about what they learned from the movies or books that Roche had recommended for them.

When he died, the funeral procession was so long that people in the rear were unable to get to the cemetery before the grave side services ended.

While working for NASA contractors in the moon program and living in central Florida, Roche and Ray supposedly met when they were working on two adjoined stages of the same rocket. By 1968 they could see that the future prospects in the space program would involve layoffs and under-employment, and began examining the idea of starting their own company.

Their efforts (like some early rockets) almost did not get off the ground.[4]

---

[4] Personal descriptions are from interviews with Chris Roche, Austin Roche, Jonathan Schmidt, and others. Resume information about Roche and Ray is from a Dun & Bradstreet report on CTC dated January 21, 1970.

Chapter 2

The Gathering: 1968

In his final interview, taped in 1986,[5] Phil Ray noted that it was clear that the space program was winding down in 1968, at least from an engineering employment viewpoint, as all the design work necessary for the moon landings had been done by then. Before leaving, he and Gus Roche were involved with the Saturn V, which was the booster rocket for the moon landings that followed.

Meanwhile, they had come across a small announcement in an electronics magazine concerning the creation of a 100-bit shift register by Philco-Ford. A shift register is basically a chip that emulates a continuous loop of magnetic tape. Under that analogy, data bytes can be written to the tape as it passes the write-head, and as the tape continually loops past a separate read-head the contents of specific spots on the tape (i.e., memory addresses) can be read back into the system. Therefore, a shift register can be used as memory for a computer. Random access memory (RAM) chips would have been much faster, since there would be no need to wait for a particular address to loop past the read-head. However, RAM chips did not then exist, and the discrete RAM circuits that did exist were very expensive. Since a byte-sized memory address (usually representing a single character) is normally 8 bits, the 100-bit chip could store only about 12 bytes, or 12 written characters. However, Ray noted, it was the first chip they had run across that handled more than a single bit of storage.[6]

"It was the first piece of what is now called large scale integration, the first large-scale chip—and no one knew what to do with it," he said. "We decided this would have a dramatic impact on the cost of technology, pushing costs down." In other words, chips would soon be inexpensive enough for business use, and cheap enough for a start-up to use them.

Having decided to leave the space program and set up their own firm, Roche and Ray called on their circle of friends for advice and introductions. One of the people they called was Victor D. Poor, then a principal of an electronics firm in Frederick, Maryland, which had recently been bought out by an electronics conglomerate, Plantronics Inc., then called Pacific Plantronics.

It was a fateful choice.

"I knew Gus for many years," Poor recalled later. "I had moved to Washington, DC, area the year before Sputnik,[7] and later we were both there working on a missile program. A couple of years later he moved on, to Florida, but we kept in contact.

"Then in 1968 I heard from Gus that they were starting a firm and needed financing. Plantronics was looking for things to expand into, besides their headset business—they

---

5 He was interviewed by newspaper reporter Richard Erickson in the late summer of 1986 as part of the San Antonio Light's multi-part history of Datapoint Corporation that ran in September of that year. Erickson said Ray appeared to be in good health at that time, but he died of cancer a year later.
6 As this is written, you can buy, in grocery stores, for a few dollars, USB key fobs that store billions of bytes.
7 Sputnik 1 was launched October 4, 1957.

had acquired my business, for instance. So I took Phil and Gus to Santa Cruz (California, headquarters of Plantronics) to meet with the chairman and president of the company. I did not listen to their pitch ahead of time, and assumed that they knew what they were doing."

He winced as they made their pitch. They said that they wanted backing so that they could acquire a company in Florida that was going bankrupt, and use it as a shell to solicit government contracts in areas of their expertise.

"There was no particular product focus—they just wanted to set up shop and get what business that they could. It was not the kind of thing that people with major capital would take an interest in. It was embarrassing. I did not realize how little either of them understood about what was needed for a business plan, or to get a business started," he said.

Ray went off that evening on other business, leaving Poor and Roche alone in the hotel.

"I read the riot act to Gus," Poor recalled. "I said that you have to have a specific product focus and market, and a believable business plan. He said, 'For instance, what?' I said, well, for instance, we have this inquiry at my firm from the Associated Press for a glass Teletype, something that will replace the old electromechanical models. I said, there is a product, and if one customer wants it, there are probably others who will want it also.

"The Associated Press wanted something that they could plug into their computers, that would put stuff on their screens, and let them retire their old electromechanical stuff. I still miss the sound of those old Teletypes. We had done a lot of business with the Associated Press, but we could not take on that project—our plate was full.

"The glass Teletype was, to me, an off-the-wall example. I did not expect him to build that particular product. I was just trying to get him thinking. But several months later I got a call out of the blue from a potential investor in San Antonio, wanting a reference on Gus and Phil. They were starting a firm, and had raised money to build a glass Teletype.

"I told the guy—Gerald Mazur—that these were good people and it sounded like an interesting project. I played the game with him.

"Then I heard no more for a while," Poor explained.[8]

Second Opinion

Roche and Ray also turned to Bob McClure, then working as a computer consultant in Dallas while teaching as well. Although younger than Ray, he had been a teaching assistant at the University of Texas when Ray was there. Ray had not only been in one of his classes, but the two had worked on an electromechanical demonstration computer, built from relays scavenged from pinball machines, that played a counting game called Nim.

"I was sitting in my office in 1968 when I got a call from Charlie Skelton, about consulting," he recalled. Skelton was another Texan and a mutual friend of McClure and Ray, who was at that time involved with Ray and Roche in their effort to start a company. "He and Phil came over and we had a discussion about computer terminals.

---

[8] The quotes are from multiple interviews and e-mail exchanges with Poor, mostly in 2008.

"One of them said they had raised money to start a new firm to build computer terminals. They had read an article in Businessweek saying that computer terminals would be the next big thing, but that neither of them had ever seen a computer terminal. By that they were joking, of course.[9]

"I said I had been thinking about the topic, as I had been doing work in computers and knew that the availability (of computer terminals) was thin. Nearly all input was done with Model 33 Teletypes at 110 baud.[10] So I said that they should build a glass Teletype. But I told them to make sure that they emulated the protocol of the Teletype exactly, so that no one at the computer end has to change any software, since that would be a hang-up to getting it accepted. There was a CRT terminal already on the market that was not precisely compatible and therefore had a lot of problems. There were some expensive machines available, but I assumed they could make one for less since there was really not that much in one," McClure recalled.

"He probably talked for two hours," Ray said in his last interview. "But the one or two sentences that stuck were the ones where he said that whatever else we do, make a product that is identical in size, and keyboard, and connections, and speed to the Model 33 Teletype."

The Business Plan

Meanwhile, Roche and Ray also appear to have taken to heart Poor's advice that they produce a believable business plan. Multiple, fragmentary versions of their plan survive in the Roche and Ray[11] family archives, mostly written long-hand and all undated. (Judging by the evidence, dating memoranda was not a common practice among CTC principals until 1972.) One included a list of computer terminal products that might be considered competitors with the one they were planning. There were 30 products on the list, ranging in price from $4,890 to $24,820. (The Datapoint 3300 later sold for about $3,500.[12] The Model 33 Teletype with a paper tape reader cost about $1,500.)

The plans are interesting both for what they say, and what they don't say. For instance, what they don't explicitly say is that, after they marketed a glass Teletype, they eventually planned to go on and create a personal desktop business computer. However, the plans contain broad hints that such was their intention.

Various sources agree that Roche and Ray shied away from using the term "computer" at the time to avoid any suspicion that they intended to compete with IBM. CTC's founders, meanwhile, were not the only ones to have this fear. For instance, mini-computer maker Digital Equipment Corp. (DEC) supposedly invented the PDP (Programmed Data Processor) nomenclature for their line of machines, starting in about 1960, to avoid calling them

---

9   Another story is that they literally did not look at any computer terminals since they wanted to approach the project without any pre-conceived notions.
10 A transmission rate of 110 baud was equivalent to 10 characters per second, or 100 words per minute, counting a word as five characters and a space. This rate was only achieved when transmitting pre-coded material from punched tape. For most protocols the characters-per-second is equal to the baud rate divided by ten, but at the 110 baud speed of the Teletype it was standard to add an extra "framing bit" for each character to give the machine's clutch sufficient time to engage.
11 After his death, Ray's archives were acquired by Jack Frassanito.
12 At any rate, analysis of early CTC financial results indicates that each unit brought in about $3,500. Then and now, vendors of industrial equipment do not publish list prices, as price is just one of many factors to be negotiated in a sales contract.

computers. Computers were monstrously expensive things that IBM made, and only a fool set out to compete with IBM. The latter's domination of the computer industry had been cemented by its 1964 introduction of the IBM System/360 general-purpose business-oriented computer system.

"Anyone who heard we were doing a general purpose computer would say we were crazy," Ray said in his final interview. When Roche and Ray started writing their business plan, the technology was not available to make a small computer, but such technology was foreseeable, he also indicated.

Meanwhile, "There is one page that clearly talks about the direction we wanted to go, and it was toward a terminal that had intelligence," he said.

While it is not certain what page of what document he was referring to in the recorded interview, one of the surviving partial hand-written business plans called for the eventual development of a "complete time-sharing system," by coming out with individual stand-alone systems that supported that goal. Of course, a time-sharing system is not complete without a computer.

Another version of the plan, one that is neatly typed and almost intact, stated that, after enhancing their basic terminal with magnetic tape storage, they planned to make a "business oriented system" with a "more extensive control unit." Presumably, "control unit" was their euphemism for computing power.

This version of the plan was 27 pages long, double-spaced, but five pages from the financial section are missing. Consequently, we don't know what the projected costs and profit margins were. Otherwise, it offers a look into their thinking, and what they thought potential backers would want to hear.

The plan lays out the advantages of computer time-sharing over individual mainframe ownership, and predicts that the market for data terminal equipment would grow from a little under $100 million in 1968 to nearly $500 million in 1971. (Ray later said that these market projections turned out to be pessimistic.) According the business plan, the overall goal of CTC was to offer this ballooning market a "low-cost, reliable data terminal."

The Product Development section of the plan called for the creation of a complete terminal in 15 months. The terminals would be sold through commissioned sales representatives, but the eventual goal was to have large companies resell them.

There is a mysterious reference to a "major supplier of telegraph terminal equipment" who was expected to place a $500,000 order for the proposed device in the near future.

The Personnel section referred to Ray as having "personally designed and directed the data communications equipment in use on almost every missile in the free world's arsenal." He was then managing a $1.3 million project to develop a satellite data communications system. Roche was described as "a foremost authority in the field of digital communication theory" and had overseen the development of systems used in the Atlantic Missile Range. Skelton, who had not dropped out of the venture yet, was described as "one of the pioneers in the application of the transistor to practical circuitry and later directed the development of the first digital computer utilizing integrated circuits."

A number of pages from the Product Plan section of another version survives, and is professionally printed. This plan presents much more elaborate product plans, and extends from the first quarter of 1969, when development of the terminal was to be finished, into 1972, when they planned to be developing flat-panel screens. (First-generation flat-panel screens had been popularized by the PLATO [Programmed Logic for Automated Teaching Operations] computer-aided instruction system at the University of Illinois in the late 1960s, and Roche and Ray presumably assumed that such displays represented the wave of the future. However, their low resolution prevented immediate widespread adoption of the technology.[13])

Before they got to flat-panel screens, the printed plan prophetically called for the use of LSI (large-scale integration, meaning hundreds rather than dozens of circuits on a chip) starting in the fourth quarter of 1970. "It will be possible, during the above time frame, to customize in LSI form most of the logic portions of the terminal systems. This work will lead to a new low-cost generation of terminal equipment in late 1971," said the plan.

Actually, as will be shown, they approached Intel about an LSI version of their processor in early 1970, but it did not appear until April 1972. As predictions go, they were amazingly accurate.

So, evidence agrees that creating a desktop computer, and the use of microchips, was firmly in mind when they went into start-up mode.

In the end, it appears that they came up with a document that would have made Poor proud with its logic and coherence, its product focus, and its reasoned assurance of eventual profits. It left open the possibility of entering the computer market, without being so presumptuous as to say it outright.

It was a fine piece of work, but there is some question as to whether it had much impact on their efforts to raise the money that launched CTC. Instead, personal connections turned out to be the key.

Mavericks

Ray had decided that what they were proposing would be too leading-edge for the staid venture capitalists on either coast, and turned his eyes to the financial "gunslingers" he had heard about in Texas. In the process, Charlie Skelton's Texas network of friends led him to San Antonio businessman Gerald "Jerry" Mazur. Mazur had gotten into life insurance after getting out of the military in World War II, and then discovered there was not much money in life insurance. Still, he met a lot of people, many with money to invest. Eventually, he became a venture capitalist on the side.[14]

First came a bowling alley, and then a series of Ramada Inn hotels, and then various business startups. Lists of backers in his enterprises included the expected mix of doctors, lawyers,

---
13 Communication with Michael Fischer, 2011.
14 Mazur was interviewed in 2008.

and community pillars, plus a scattering of Mavericks.[15]

When he was approached by Skelton, Roche, and Ray, any caginess they had about using the word "computer" was completely lost on him.

"I understood that it was supposed to be a computer firm," Mazur insisted, four decades later. "They probably figured I would not know what a terminal was. I am sure they called it a computer. I said to them that I know nothing about computers and that you don't need me, and they said they did since I was a businessman."

As for the technology, "With most of the firms I have been involved with, I did not know much about it, but I think I know about people. I limited myself to dealing with people and not technology. When I met Phil and Gus I liked them. I thought they were top-flight scientists," he recalled.

Investors were recruited mostly via word of mouth, he recalled, starting with people who had invested in his previous ventures. All the money was raised in San Antonio.

"I'm a hell of a good salesman, and there were not a hell of a lot of people that turned me down," Mazur explained. "I would say, 'You don't want to invest? This is the opportunity of a lifetime. I thought you had more perspective than that.' And then I would get up to leave, and they would say, 'Wait, I'll put some money with you.'

"You use centers of influence," he added. He once explained to Jack Frassanito (who joined the company after it was formed—see the next section) that he had learned to raise money by fishing, where he found he should not hit the fish on the head with the lure. The angler, he learned, should place the lure in front of the fish and then pull it away slowly. Seeing that it might lose the supposed morsel, the fish will begin following the lure, and will eventually swallow it. Investors often respond to opportunities in the same way, he said.

Indeed, speaking of fishing, one wealthy accountant offered to invest, but only if the sportsman-businessman who had an office on the next floor went in first. So Mazur went upstairs.

"He had a big blue marlin on the wall, and I mainly sold him by telling him how much I admired him for being able to catch it. He bragged about other catches and I won him over. And I came downstairs and said (to the accountant), 'You just made an investment.'"

In the end he raised nearly $650,000, all in San Antonio, apparently in the course of a few weeks in the spring of 1968. (According to one story, Roche and Ray figured they needed $300,000, and Mazur, based on his experience with start-ups, doubled that and added $50,000.)[16] About half came from one individual: Joe Frost Jr., a director at Frost Bank, a leading financial institution in San Antonio, founded there in 1868. (His cousin Tom Frost was head of the bank, but Mazur recalled that Tom Frost was one of the few people who turned him down.)

---

15 The uppercase M is not a mistake. In San Antonio, a Maverick is a descendent of Sam Maverick (1803-1870), the San Antonio lawyer and real estate developer who inspired the dictionary entry. However, a final list of CTC start-up backers has not surfaced.

16 Datapoint vanity history of 1982.

"We were judged by the look in our eyes," Ray recalled in 1986. "(Joe) Frost did not even listen to the technical part of the presentation." Frost's son also loaned Ray money so that Ray could invest in the venture, as having large investments from founders looked good to the other investors.

Common wisdom is that CTC was set up in San Antonio because Frost insisted that his investment remain in the city, for the city's good. Frost evidently endorsed the idea, but Mazur said it started with him.

"I required that firm be located in San Antonio, not Joe Frost," he said. "They wanted to go to Dallas where plane connections were better, but I said I would not do this thing unless they stayed. I was not going to move out of San Antonio, I think San Antonio is heaven."

Others did not share Mazur's feelings about the city, and Skelton dropped out, reportedly after his wife declined to move there. (Recruiting people to move to San Antonio, which for many people was equivalent to Timbuktu, was an on-going problem. Ray was able to get one man to make the move after convincing the recruit's wife that Texas was not, as a she had heard, crawling with revolting bugs. For months after arrival, every day she would send Ray another jar containing yet another horrid bug she had found in her new home.)[17]

Moving was also wrenching for the Roche family, but they went through with it. "It was presented as an adventure," recalled Chris Roche, who was in sixth grade at the time. His father had to choose between a new job with Hewlett-Packard in California, or starting a business in Texas, but either way they would have to move.

Back in Florida, the family often spent weekends snorkeling in lakes while their father went cave-diving below with a particular fireman as his sole, trusted diving partner. Fishing, target shooting, and camping on the beach were also on the weekend agenda. Gus Roche was in the local hot rod crowd, which included some astronauts, and had at times a Triumph TR3, a Mustang GT, and later a Porsche. At times when he was out of school, Chris Roche recalled going with his father on special trips to Cape Canaveral to install equipment.

"It was an interesting time in Florida—troops were billeted in our school playground during the Cuban Missile Crisis. We would watch (NASA) rocket launchings on TV and then step outside and see them in the sky," Chris Roche recalled.

"We had a bohemian lifestyle in Florida, and lived on the beach almost every weekend in the summer, where it was cheaper to catch crabs and have cookouts than eat at home," recalled his younger brother, Austin Roche. "When we heard the decision to move to Texas, we were devastated. Why would we want to go from lush green Florida to dry hard Texas? Then dad would tease us and say that when we got there the first thing we'll have to do is get guns."

With a small staff, Roche and Ray set up operations in a rented building with about 2,000 square feet at 142 West Rhapsody, in a light industrial district just south of the western edge of the east-west runway of the San Antonio International Airport.[18] The northern edge of the developed area of the city was just north of the airport at that time, although four decades

---

17 Ibid.
18 The building and the district are still extant. At this writing the building appeared to house several small contracting firms.

later it had spread another dozen miles northward.

CTC first incorporated under Texan law on July 6, 1968, with six million shares outstanding. Its fiscal year began July 31.

Their staff was small, as Roche and Ray did not need much help with the electronics. One of the first persons they hired was Richard "Dick" O'Connor Norman, who also came from Florida, where he had worked for Martin Marietta.[19] He brought several other engineers with him, men who would later follow him to other jobs. Norman would handle problems posed by actually producing CTC's (and later Datapoint's) products.

He would be with Roche literally in Roche's last hour.

The Designer

One of the first decisions that Roche and Ray made about the product they were planning was that it was not going to be built by engineers for engineers. It needed to look nice.

"My brother (Burt) is an architect[20] and when we were putting together this document (the business plan) I asked him to sketch out what a terminal should look like and he was not interested in doing this. He said he did not have the time and had had enough of his brother," Ray said. "I was disappointed, and I decided, I suppose probably out of sheer vengeance, that I would go to the best designer on earth, that is Raymond Loewy. I wrote him a blind letter and got a reply that they were very interested."

Ray's archive included the carbon of a typed letter dated August 23, 1968, to Raymond Loewy's firm at 425 Park Ave. New York City. It stated the CTC had recently been organized and needed an industrial designer, and asked for information concerning capabilities and pricing. (Ray actually wrote to other design firms as well, apparently without result.)

Whatever his motivation, Ray was shooting for the big time with that simple letter. Raymond Loewy (1893-1986) was the premiere industrial designer of the twentieth century, and can be said to have founded the discipline. Signature designs included the Studebaker Avanti, the 1950s version of the Coca-Cola bottle, the Lucky Strike cigarette pack, the Greyhound bus, and the Shell logo. He was behind the appearance of so many corporate logos and household appliances that the average American probably spent most waking hours with at least one of his designs within sight.

Dozens of designers worked for Loewy, in offices in the US and Europe. He sent out John "Jack" Frassanito, who had joined the firm only that March and was working in Loewy's New York office.

Then 26 years old, Frassanito had started out in the auto body business in New York, but had an epiphany after taking elaborate pains to repair a turquoise 1957 Chevy. The owner agreed that it looked like the day he had bought it—and Frassanito decided he did not want to spend his life in situations where the best work he could do was to make something look like it had never been damaged. He decided to become a car designer, and applied to the Art

---

19 Details on Norman are from a 2009 interview with his daughter, Chrisa Norman Scoggins.
20 He was involved with the design of the Institute of Texas Culture, a museum in San Antonio.

Center College of Design, then located in Los Angeles.

They rejected him.

So he sold his share of his co-owned repair shop, packed his belongings in the back seat of his Stingray, and drove to California, where he confronted Carla Martel, the director of admissions. He said he had to get out of the car repair business. He said he was tired of arguing with people, of driving tow trucks in the middle of the night, and getting dirty.

She got out his records. "You don't have any talent and your grades are terrible," she noted. There was an awkward silence. "And you knew that? And yet you drove all the way from New York to make an appeal?" she continued. "Well, we'll let you in."[21]

The school had three semesters a year, and Frassanito got a bachelor of science in two years eight months. The schedule was brutal, with a class every day and a project based on that class due a week later. Out of 40 freshmen, four graduated, he recalled later. Most who survived the first year got scholarships, and Frassanito got one from Chrysler, and later a job offer from both General Motors and Loewy. The latter, however, was involved with Skylab. Having been an avid childhood follower of the Disney/von Braun space exploration articles in Colliers Magazine, Skylab sounded more exciting to Frassanito. So he went back east and found himself working with von Braun and the Loewy Skylab team, fulfilling a childhood dream.

By the end of the summer he had been involved in the preliminary decision that the lab ought to be launched dry (not full of hydrogen fuel, since it was a converted fuel tank) with the furniture already installed. But he could also see that the project would not come to fruition for years. (In fact, it was not launched until 1973.)

As a newcomer with Loewy's firm, Frassanito was sent to Texas for the CTC project, which was not considered a plum assignment, he recalled.

Frassanito flew to San Antonio and stayed in the La Quinta Inn near the airport. On West Rhapsody, he found about two dozen people working in the cramped space. The Datapoint[22] 3300, as they named their glass Teletype (since it was supposed to be 100 times better than the Model 33 Teletype) did not even exist yet as a sketch, he recalled. Phil Ray had come up with the Datapoint name, he was told, while being driven to the airport one day. He had generated a list of syllables that could be used to produce a short word that ended with a hard consonant. This approach had proven successful in names like Xerox, Kleenex, Rolex, and Ajax. He stopped after combining data and point

Frassanito commuted back and forth to New York (where he was still involved in Skylab) every other week. He was pleasantly surprised that San Antonio did not look like the surface of the moon, as he had envisioned. It wasn't New York, but it was pleasant enough. Soon he was staying with the Roche family at their house in Alamo Heights, an old-money bedroom community with its own city government in the shaded hills just northeast of San Antonio's downtown. (It was famous for having a garbage service that did not require citizens to put their refuse on the street for collection—the collectors went into each backyard to find and

---

21 Quotes from Frassanito are from interviews and e-mail exchanges, mostly in 2008 and 2009.
22 For reasons now lost, the one-word version, Datapoint, soon won out over Data Point.

remove the garbage.)

He discovered that he and Roche were both sportsman. (Frassanito did skydiving as a hobby, while Roche had taken up sailing.) Roche was also intrigued by Frassanito's ability to draw, and the possibilities this opened up in terms of communicating his ideas. Roche himself tended to get less coherent as he got more excited about a topic.[23] Frassanito accompanied the Roche family on various vacations and sailing trips, and found himself drawn into lengthy discussions with Gus Roche about anything and everything—including the future of computing. However, only about a quarter of their conversations involved computers, he later recalled.

For instance, Roche was very concerned about how groupthink affected organizations, but when discussing the topic he found it easier, out of frustration, to make allusions to various classics. These Frassanito ended up reading so he could keep up. References included "The Trial" and "The Castle" by Franz Kafka, Gibbon's "Decline and Fall of the Roman Empire," and Machiavelli's "The Prince" and "Discourses on Livy." Frassanito decided he liked Machiavelli's "Discourses" more than the better-known "Prince."

"Gus was talking about the absurdities of bureaucracies and how they behave, and how people behave and act in bureaucracies—how they can be irrational and self destructive. How they tend to position themselves and resent each other. And how companies usually come apart from the inside out rather than the outside in," Frassanito recalled. Kafka and Machiavelli were drawn on for examples of how aggression and arrogance can tear an organization apart. Gibbon was seen as an authority on how major institutions come and go.

When the conversation turned to technology, Frassanito realized that Roche was building the Datapoint 3300 glass Teletype mostly as a ruse. "Gus said it was just a temporary product to get the company started, with something that was as technologically simple and straightforward as possible," Frassanito recalled. "He said that the Datapoint 3300 was a flash in the pan—its market was not that big, and anyone could duplicate it rather quickly since it involved no technical advances.

"But Gus said he knew a personal computer was coming down the track—he had heard university professors talking about using intelligent machines, and he wanted to be there on the day that TTL[24] enabled intelligence to fit into a machine that can sit on a desktop and communicate with the outside world," Frassanito explained. "If you told potential backers that you are going to make a computer on a desktop, you would get laughed at. If you told them you were going to compete with IBM, you would get chased out of the room. The original business plan to start the firm and produce the 3300 was intentionally couched in terms that would retire the fears for non-technical people that you would be trying to do something that could not be done.

"There needed to be another product, or the company would go away," he recalled Roche saying.

---

23 Frassanito liked to compare the results to Jack Lemmon's portrayal of an agitated nuclear power plant shift supervisor in the 1979 movie, "The China Syndrome."
24 TTL refers to Transistor-Transistor Logic, at the time the most widely used and cost-effective technology for integrated circuit chips that embodied a significant number of devices, or logic gates. Previously, the industry relied on much larger (although often faster) discrete transistors, resulting in room-sized computers.

Basically, the philosophical discussions went on for a while, and then they started talking about the configuration of the next product. These involved late-night brainstorming sessions at San Antonio's University Club, during which Frassanito would wince while Roche drew diagrams on perfectly nice white tablecloths. On the other hand, the management of the elegant restaurant never complained, probably because they were mindful of the amount spent there from CTC expense accounts (especially the bar tabs.)

During one such late-night session, Frassanito asked how an "old fellow" like Roche managed to work so hard. Roche noted that he got his CTC stock at ten cents per share, but when it went public its worth—thanks to his hard work—rose by a factor of 80 to $8 a share. Frassanito immediately saw the advantage of being an entrepreneur.

McClure's dictum to keep their product plug-compatible with an existing product seemed to have worked well, so they cast about for another widely owned electromechanical device that was ripe for replacement with electronics.

The apparent answer was the IBM 029 keypunch machine. Computers were programmed by feeding them stacks of punched cards. Each card contained one line of programming code, or data, with a maximum of 80 characters. The contents of the card would be typed across the top, and each letter or number in the header would be encoded by holes punched in the columns below it. Naturally, every place that used card-reading computers needed punch-card machines, especially as they were the standard way to input data, such as payroll time sheets. An electronic version of the 029 would store the data on tape for transfer to the mainframe (or transmit it via a phone modem) rather than generate stacks of punched cards. ("Plug-compatible" meant that the data reaching the mainframe from an electronic 029 replacement would be indistinguishable from data generated by an original 029, as there was no actual plug connecting a 029 to a mainframe.)

CTC's engineers pointed out that a hard-wired emulation of the 029 could be gotten to market rather quickly, but Roche thought it was now possible to put a general-purpose computer inside the machine, which could be programmed to perform the desired functions.

But first, they had to get the company launched by selling the Datapoint 3300.

Chapter 3

Tooling the Datapoint 3300: First Half of 1969

At the same time that Frassanito was socializing with the Roche family and exploring philosophy and computer design with Gus Roche, he and the rest of CTC were working hard to bring the Datapoint 3300 to life.

Work began in earnest in the late fall of 1968. Looming on the horizon was the 1969 Spring Joint Computer Conference, to be held May 14-16 in Boston. They wanted to debut the new product there, as it was then the premiere gathering of the fledgling computer industry. Wall Street bankers were known to prowl the aisles to see what the latest gadgets were—and decide which to invest in. Knowing they would eventually need more money, Roche and Ray wanted to be seen there. Of course, attending would be expensive, and arrangements would have to be made well in advance. Spending that money with nothing to show for it could have ruined a small start-up like theirs.

Chris Roche (Roche's eldest son) recalled circuit breadboards, with spaghetti-like tangles of patch wires, laid down end-to-end on folding tables, snaking through the office on West Rhapsody. Chips would come in and be added to the assembly, replacing some of the breadboards and shortening the line of tables.

Indeed, according to Ray, the designers were counting on new chips from Texas Instruments that were essential if the machine were to work as advertised.

"I had worked for TI right out of school, and still knew some people there when we started to design the 3300," he recalled. "We went to TI and asked what is on the drawing board, and asked if they could make a chip that could contain all the characters of the alphabet. They said they were working on it, and we said we wanted the first one. But when it heated up it started to lose characters."

TI promised an improved version that would work reliably. Meanwhile, acquiring components from other suppliers where CTC's founders did not have contacts involved subtle problems. People would ask, "Where in California is San Antonio located?" Told that it was in Texas, they would then ask if it was accessible to a major city with an airport.[25] Typically they had heard of Austin (which was actually one-third the size of San Antonio) and knew that Austin had an airport. San Antonio, of course, did have an airport, and the CTC staff could hear the planes coming and going two blocks away.

At one point Ray put a call to the Dallas office of a supplier, and after explaining that he was interested in buying large quantities of integrated circuits, the salesman put him on hold for an extended period. Later, after getting to know the guy, Ray found that the salesman had spent that time laughing hilariously and telling his co-workers about this crackpot in San Antonio who said he had started a computer company. His co-workers urged him to humor the lunatic, since he claimed to have funding, and so the call was resumed.[26]

---
25 The author had similar experiences with callers while working for Datapoint 12 years later.
26 The anecdote is from the unpublished Datapoint vanity history. The city has enjoyed more national mindshare since the arrival of a professional basketball team, the San Antonio Spurs, in 1973.

Enter Intel

They had no such problems with a fellow startup that actually was in California, called Intel Corp. In fact, this firm—fated to play a large role in this story—had been started at almost the same time as CTC, incorporating on July 18, 1968 (originally under the name NM Electronics.) It originally had 12 employees and revenue for the first year was $2,672.[27] Obviously, they could not afford to hire salespeople—but they still managed to connect with CTC, halfway across the country.

Most of the founders of Intel had been veterans of Fairchild Semiconductor. Bob McDowell had been with Fairchild in Dallas, and had left, hoping to become a salesman for the newly founded Intel. However, he recalled four decades later, they were only able to make him a "manufacturing rep," meaning he became a freelance salesman who represented Intel as well as other electronics firms.

"When I joined, their inventory was in one file cabinet drawer," he said. "At their first sales meeting they announced a product they would second-source for Fairchild, a 256-bit shift register."

Second-source arrangements sprang from the fact that makers of end-user electrical equipment are never thrilled about designing components into their products that could only be acquired from one vendor. Consequently, component makers have found it advisable to foster competition and license other vendors to make their products. As previously explained, shift registers were the form of memory used before RAM chips came into wide circulation. They emulated a loop of magnetic tape, where sequential memory addresses, with their contents, rotate past a read-head.

After leaving the meeting at Intel, McDowell got a call from Ray at CTC inquiring about another vendor that McDowell represented, a maker of keyboards. So he went to San Antonio, and met with Ray about keyboards. It was during the 1968 Christmas season, he remembered.

"At the end of the conversation I mentioned coming back from a startup in California called Intel and showed him the data sheet for the shift register. At the time, 256 bits was big," he recalled.

"Phil suggested that I should be wearing a Santa suit, and said, 'Damn, can these guys ship this? Can they meet National's price?' I got on the phone to Intel, and they came unglued. We met the price. I wrote a million dollar sales contract that very day. We started a relationship with them (CTC) right then."

McDowell later surmised, from comments from others in CTC, that National Semiconductor was withholding components from CTC, saying the items were "on allocation" (i.e., being rationed to the buyers due to a possibly artificial shortage) but that CTC might get more favorable treatment if National got some of CTC's TTL component business that was going to Texas Instruments. Intel's timely arrival gave them leverage against National—and explained Ray's reference to Santa Claus.

---

27 Details about Intel are from its 1984 16-year anniversary brochure, "A Revolution in Progress."

To cement the relationship with Intel, McDowell suggested that, in exchange for being the sole source of the shift registers, Intel inventory its components inside CTC's facilities. Bulk deliveries were made to CTC, but Intel only charged for what CTC used in a given week. McDowell went there weekly to verify consumption and prepare the invoices.

The Show Goes On

Meanwhile, the tangle of wires that was strung across multiple card tables was first made to function as a terminal, putting characters on a screen, on January 2, 1969. Ray took a picture of the screen to commemorate the event. But, after 30 minutes, heat stress left the screen blank. A reliable character generator chip arrived from Texas Instruments only days before the May conference, Ray recalled.

While the electronics were being made to work, Frassanito was struggling with the terminal's enclosure, or cabinet. The mold used to make the ABS plastic version of the cabinet would not be ready in time for the impending show in Boston. He knew a model shop in California he had worked at while in school that could produce fiberglass prototypes that would be suitable substitutes, and prepared to rush out there and spend a final weekend, and get them made.

Then Loewy called and told him to return to New York and the Skylab project, which, to them, had a higher priority.

Not getting those prototypes made would mean that CTC would have nothing to display at the show, and as a result would probably not survive. Roche and Ray had become his friends, and Frassanito did not want to let them down. And there was another thing: "I had already been talking with Gus about designing a personal computer," he said.

So he went to California and worked around the clock and finished the prototype enclosures in time—and burned his bridges with Loewy. He soon became a full-time CTC employee—at a higher salary.

Roche, Ray, and Frassanito flew to Boston with the three hand-finished prototypes, buckled into first class seats—they were too important to entrust to the escorted luggage service. The three executives rode with them.

The 1969 Spring Joint Computer Conference was the place to be. According to coverage in the New York Times,[28] 170 companies exhibited, and another 50 were turned down because of lack of space. The overflow first-day crowd packed the aisles until they were nearly impassable, and there were an estimated 20,000 visitors during the show. (Four decades later, the premiere high-tech convention is the Consumer Electronics Show, currently held each January in Las Vegas. It draws ten times that number of exhibitors and attendees.) The NY Times correspondent marveled at the youth of exhibitors, two-thirds of who appeared to be under 45, and a third appeared to be under 30. (The show was sponsored by the American Federation of Information Processing Societies, an organization that has been defunct since 1990.)

---

28 New York Times, May 15, 1969, page 74, "Computer is Called World Economy Aid."

Meanwhile, Wall Street scouts were, as previously noted, indeed prowling the aisles looking for likely ventures to invest in. One of them was John Bender, then 24, a graduate student at Case Western Reserve who was there specifically to examine CTC from a technical perspective for an investment banker who was interested in the startup. Bender, the banker, and another student ended up at the CTC hospitality suite at a nearby hotel for an interview with Roche and Ray.

Bender recalled that the suite was sparse. There were no slick displays, carts of finger food, show girl hostesses, or beverage bars with uniformed bartenders, as would become obligatory in later decades at trade shows in Las Vegas. There was, however, a working prototype of the Datapoint 3300.

Bender recalled talking to Roche for a few minutes and then having a long conversation with Ray. Both men seemed quick-witted and self-assured, and exhibited an impressive level of intensity, although Ray, with his Texan persona, seemed a little less intense than Roche.

"They were both jazzed up and excited about that they were doing—it was a high-energy experience," he recalled.

In his conversation with Bender, Ray went into considerable technical detail about the 3300—and then branched into the CTC's future intentions. These, Bender soon saw, involved a one-bit desktop computer.

First, Bender expressed concern about the 3300's display, which flickered more than he liked. He discussed persistent phosphors that CTC might use, and found that Ray knew more about the topic than he did.

"IBM terminals were horrible for flickering, and if they got a little out of adjustment waves would run across the screen. At least the 3300 didn't have that problem," Bender said, adding that the flicker on the 3300 was cleared up before it got into production.

Discussion then turned to the computational power represented by the TI character generator chip. Ray said that time would soon come when a processor on a chip could do more than just generate characters and run the display. It would be a programmable computer. The question was not if it would happen, but when, Bender recalled Ray saying.

Ray suggested that the upcoming processor-on-a-chip would have a data path one bit wide. "I had not thought of that before and was skeptical about what could be done with a one-bit processor. We had a discussion about how you could do things with one bit.

"We talked about teaching terminals, since that was the area of my dissertation, and how much smarts you could put into a terminal to take the load off the mainframe, and how much processing you could do in the terminal versus the mainframe," Bender said. "Mainframes at the time were slow when serving multiple terminals, so the more you could offload the better."

After the meeting, Bender compared notes with the other two. The investment banker saw CTC as a computer firm that had the potential to become the next DEC. There seemed to be no doubt that CTC's next move would be to develop a desktop computer. Bender and the other student were both enthusiastic about what they had seen.

"We saw it as the way things were going to go—if not these guys, then someone else would be doing this," Bender recalled. The investment banker did recommend that his firm invest in CTC, but there was never any follow-up.

Back at the show, the CTC booth was small and near the back of the hall—a prescription for being ignored. So, recalled Ray in his final interview, Roche and Ray took their terminals to the booths of computer vendors, unplugged the clattering Teletypes with their spools of paper, and plugged in the silent, paperless Datapoint 3300s. They worked without a glitch, and, Ray recalled, "They were the hit of the show."

High-Tech Elegance

Why did they make a big deal about the Datapoint 3300? Decades later, Jonathan Schmidt, who joined CTC near the end of 1969 (i.e., after the 3300 was already on the market) still remembered his first encounter with it.

"I was blown away by its elegance—it was like a Ferrari when all you have known was a '49 Ford," Schmidt explained. "I had never previously encountered a piece of electronics that was so beautiful or worked so well. Any other CRT console was sheet metal with punched holes, and looked heavy and clunky, and nothing on it looked quite straight—like a bad mock-up from a science fiction movie," he said.[29]

Also, the appearance of the text on the screen of the Datapoint 3300 was crisp, and the screen could display 72 characters on a line, the same number as a Model 33 Teletype could print on a line of paper. Other glass Teletypes typically used large, fuzzy, jittery characters, and often could only display 40 characters on a line—or almost any number besides 72—creating immediate incompatibility with the Model 33.[30]

The base of the 3300 was designed to fit in the desktop "footprint" of a Model 33 Teletype, so the buyer would lift the Teletype off the table and put it aside, the put the Datapoint 3300 down in the same spot.

Schmidt recalled that Frassanito had surveyed office workers and found that their favorite typewriter was the IBM Selectric. Consequently, the slope and layout of the keyboard on the Datapoint 3300 was modeled after the Selectric.

The biggest advance, as far as Schmidt was concerned, was the machine's use of "keyboard rollover." Basically, the Model 33 Teletype did not have rollover, and pressing one key locked the other keys. A fast typist who wanted to input THE might press T and then immediately H, but the H would be locked until the T was fully released, which would typically not happen until about the time the typist pressed the E. So the typist ended up with TE, if not T. Consequently, a typist was limited to a halting, two-finger speed.

With the Selectric, however, you would run your finger across the keyboard, like a glissando on a piano, and it would print every letter whose key was depressed, in the correct order. Consequently, you could type as fast as humanly possible and the machine would faithfully

---

29 Schmidt was interviewed in 2008.
30 How CTC accomplished this was the subject of U.S. Patent 3,706,905, applied for on May 14, 1970, and granted on December 19, 1972.

record each keystroke. The designers of the Datapoint 3300 managed to achieve much the same thing, and this had never previously been done with a computer terminal, Schmidt recalled.

Of course, being electronic it was essentially noiseless, unlike the clattering Teletype, and it did not consume rolls of paper and spools of inked ribbons. Linkages did not break and gears did not wear out. Since it was not constrained by the speed of its mechanical parts, it would offer much higher transmission speeds than the Teletype's 10 characters per second. If a proper connection was available, it could (optionally) go as fast as 480 characters per second—a dizzying speed at the time.

The Lease Paradox

By the time of the Boston computer conference, the original $650,000 was used up. Roche and Ray were able to go to Joe Frost Jr. and show him some initial orders, and he agreed to loan them an additional $150,000—with conditions. The first condition was that Gerald Mazur take a greater role in running the company. The second was that they take the company public.

The loan let them expand to 88 full-time employees, of whom 55 were on the production line.

In October of 1969, just after they began shipping the Datapoint 3300 for revenue, CTC raised $3,655,000 with an initial public stock offering of 400,000 shares on the over-the-counter market. Of that, $800,000 was used to repay a loan from an unnamed shareholder, $478,000 was used to buy equipment, $450,000 was used to acquire the building and land that became the firm's new headquarters in northwest San Antonio, and the rest was treated as operating capital. The building was custom-built and leased to CTC with a purchase option. The story is that the builder figured the place could always be used as a bowling alley, if CTC failed.

Meanwhile, thanks to their successful launch in Boston, Roche and Ray soon got more orders than they literally knew what to do with, since mass production of the cabinet was still not ready. So they contracted with a local factory that made motorcycle helmets, to make fiberglass versions of the enclosure. They figured, correctly, that anyone who could handle the smooth curves of a helmet could handle the smooth curves and corners that were the signature of the Datapoint 3300. The units had to be expensively finished by hand, with sanding and painting, but they decided to bear the cost, since the aesthetic appearance of the cabinet was an important part of the product. Anyway, to change the appearance would be tantamount to coming out with a new product.

By the end of July they already had orders on the books for 876 terminals, worth $2.9 million.

Shipment for revenue started on September 21, by which time the number of orders had risen to 1,405 terminals, plus 141 magnetic tape decks. (The latter were meant to replace the paper tapes that Teletypes used to record text for subsequent retransmission. Ray later noted that there was really not much demand for them.) The orders were worth $5,093,345.

Since the money raised by CTC to that date was less than the orders on the books, the firm appeared to be already profitable, after operating for hardly a year.

Actually, those revenue numbers were a bookkeeping fiction. In reality CTC was facing more than two years of financial misery and related management turmoil.

Notice that the sales were described as being "worth" a certain amount. That does not mean they actually brought in that amount—at least not immediately. But, unfortunately, the money was needed immediately.

Basically, CTC (and everyone else in the computer industry) had to model their marketing practices after those of IBM, since corporate executives who bought computers were accustomed to nothing else. A big reason for that is that many of corporate executives had previously worked for IBM, Big Blue being the only significant source of experienced computer managers at that time. IBM was said to be more than philosophical about seeing them lured away by customers, and wags referred to the resulting class of IBM alumni as The Brotherhood of Men in White Shirts. They could be expected to buy IBM products for their new employers, unless there was a compelling reason not to. But, troublesome as that situation sounds, it wasn't CTC's primary problem. The problem was that IBM did not sell products—it leased them. Consequently, CTC had to lease its products, too.

For IBM, leasing was a diabolically clever strategy. For its competitors, it meant flirting with bankruptcy.

As Ray later explained, if you had a product that cost you $1,000 to make, you could count on being successful if the market was willing to pay $3,000 for it. But if you had to lease it instead of sell it, you could charge maybe $100 per month and you would not be profitable for nearly a year. After that, any additional lease revenue was gravy, but would you survive long enough to enjoy it? Each unit you manufactured in the meantime would involve additional expenses that would not be recovered for nearly a year. Therefore, as your business expanded you were actually digging yourself a deeper financial hole. For CTC, the only short-term relief was occasional cash sales to resellers.

IBM, meanwhile, was smugly immune to the lease paradox, thanks to its broad product line and cash reserves. Leasing was, however, a brake on innovation, as there was no compelling reason to produce new products when the old ones took so long to pay for themselves—and thereafter remained cash cows. But there was no guarantee that IBM would remain asleep, and always in the back of the minds of CTC's management was the fear that Big Blue would bestir itself and unveil something that would make CTC's entire product line obsolete overnight. (Those leases, after all, had cancellation clauses.) Radically advanced technology would not be required—some must-have widget that was incompatible with CTC's products would suffice. CTC executives would cringe when, at trade shows, they would see IBM executives eagerly examining the 2200. (As will be seen, IBM finally made its move in 1981.)

CTC did eventually reach a "cross-over point," in mid-1972, where it could live off its lease revenues. In the meantime the lease paradox triggered considerable management dissension within CTC, with some insisting that expansion be slowed, and others insisting that they seek further investment, although this would dilute the ownership of the original investors.

Fast Cars, Big Losses

What Mazur was still bitter and disillusioned about four decades later was the Porsche incident. Various sources tell the story in slightly different ways, and no one can pin a date on it, but it must have happened after the largesse of the IPO. Apparently Mazur at some point rebuked Ray for driving a beat up old car to work. Having a rust bucket in the president's parking space was not good for the corporate image, etc. He suggested that Ray buy a new car with company money. Perhaps to take the sting out, he suggested that Roche and himself should do the same.

Supposedly, Ray said he would buy a two-door, six-cylinder compact car—in other words, something corporate. Mazur was agreeable.

But the car that Mazur had maligned as an eyesore was Ray's old Porsche Speedster convertible, which Ray loved despite its ragged top, and even its lack of air conditioning to counter the Texas heat. So Ray ran out and bought a silver Porsche 911, which was indeed a two-door, six-cylinder compact, but also, at $12,000, was said to be one of the most expensive personal cars then on the market.

Nothing loath, a few days later Roche acquired a bright blue Porsche 911, also with company money.[31]

Mazur was beside himself with fury. He confronted Ray, who pointed out that it was all Mazur's idea. This failed to mollify Mazur, who rushed downtown to see Joe Frost Jr., demanding that Roche and Ray be fired.

Frost proved to be more philosophical, and pointed out that Roche and Ray were indispensable, implying that their retention might be worth a Porsche or two. He told Mazur to institute a written policy concerning company cars, with cost limits. Mazur did, and the uproar died down. Hard feelings, however, remained, as did the dissension arising from the lease paradox and the daily burn rate—the amount of money being lost daily—that it produced.

Ray would later say that the burn rate amounted to five Porsches ($60,000) per day. That was apparently an exaggeration, according to the averages shown by the annual reports—but that's not saying that the burn rate could not have peaked at that sum on a bad day. Assuming 260 business days per year, the burn rate amounted to $4,679 per day in fiscal 1970, when annual losses amounted to $1,216,557. It exceeded a Porsche daily in fiscal 1971, when the annual loss of $3,749,969 produced a daily burn rate of $14,423. For fiscal 1972, when losses totaled $2,220,000, the burn rate fell to $8,538 daily. Profitability arrived with fiscal 1973.

Throughout all their travails, CTC's management remained aware that they couldn't count on the Datapoint 3300 being a successful product indefinitely. There were at most 150,000 electromechanical Teletypes eligible for replacement with a glass Teletype, so the potential market was hardly unlimited. Even if there wasn't, there was nothing about the technology

---

31 Observing that the sight of an engineer in a Porsche did not always sit well with Wall Street types, Roche later acquired a conventional station wagon with simulated wood siding, principally to ferry visiting financiers to and from the airport.

of the Datapoint 3300 that someone else couldn't mimic.

As noted in the previous chapter, Roche and Frassanito had a follow-on product firmly in mind. They pursued it despite pushback and financial uncertainty, and the fact that it had never really been done before. In fact, there was no real name for what they wanted to do, and they settled on inadequate and ultimately misleading descriptions, like "intelligent terminal" or "programmable terminal."

As Gutenberg had invented the personal book, CTC was about to invent the personal computer.

Chapter 4

Inventing the Personal Computer: Second Half of 1969

As sales began to ramp up for the Datapoint 3300, two previously mentioned facts took on increasing importance for the management of CTC, recalled Frassanito.

The first was that plug-compatibility was the way to go. Making a product that could be swapped with a Model 33 Teletype without any computer modifications or reprogramming had proven to be a hit.

"McClure's bit of insight worked like a charm," he said. "His concept that it should be as simple as possible to install sounds simple now, but it was not in those days."

The second self-evident fact was that the Datapoint 3300 would be a flash in the pan, and that they needed to start coming out with a second product. Frassanito recalled that his philosophical discussions with Gus Roche often turned to what the next product should be.

"As Gus and I talked about the next thing down the road, we saw that we needed to find a ubiquitous product that we could play the plug-compatible game with. The best candidate was the IBM 029 keypunch. There were millions of them out there. We could unplug the 029 and plug in our machine—it had worked so well with the Datapoint 3300. We would use the 029 as a baseline for what our machine would do," he said. (Plug compatibility was actually easier with the 029, since its only direct connection was to the power outlet.)

Making the screen of the Datapoint 3300 the same width as the printed line of a Model 33 Teletype have been one key to its success, so it seemed obvious that the new machine should have a screen that emulated the layout of an IBM punch card, which had 80 columns and 12 rows. That would make it slightly wider but half the height of the screen of the 3300 (and most other computer displays in the industry) and so its cathode ray tubes had to be custom-made.

Meanwhile, at the time, remote data entry often involved people punching information on computer cards and mailing them to corporate headquarters, or to their computer service bureau, where the cards were fed into a mainframe. Even the slightest error would cause the mainframe to spit the cards back out, and the process would have to be repeated, imposing expensive, irritating delays.

Consequently, they decided it would be a good idea to give their machine the ability to validate or verify the information before it was sent in. For instance, fields that were supposed to contain numbers should not contain letters, payroll dates should not be a century in the future, and a week cannot have more than 168 hours.

"We wanted to replace punch cards with magnetic tape and an acoustic telephone modem and send conclusions rather than data," Frassanito recalled.

Of course, such abilities implied that the machine would have some intelligence and could therefore be used for a wide variety of general tasks. But discarding the mainframe and doing all data processing locally, on the new machine, was not initially considered an option for the 029 replacement, he added.

"We were using what existed. We would not go in and say, 'The day has come when you don't need that IBM mainframe anymore.' That was a nonstarter."

Eventually, they did actually replace mainframes. But at the time, the whole project was almost a nonstarter.

Pushback from the Staff

"The first guy who didn't want to do it was Phil Ray," Frassanito recalled. "He had made some money and was flying high, and the last thing he wanted to go through was another product launch. The rest of the company felt the same way. Gus said that we had a year to come up with a new product, and if we did nothing then all the income would go away. After sleeping on it for a couple of days, Ray came around," he recalled.

A bigger stumbling block was encountered when they sat down with the Datapoint 3300 design team.

"Sitting around a table with the original design team, we told them that we were going to do another product, and here is the concept—we have decided that we want an 029 plug-compatible keypunch. They said that's easy—they could reverse-engineer the 029 and hard-wire it in TTL and make it plug-compatible, like we did with the 3300.

"But our original idea was to have a processor. They could have gotten to market faster with a hard-wired product, and I was the only guy in the room who thought it was a bad idea," he recalled. (A hard-wired product would have mimicked the functions of the electromechanical parts of the 029 with electronic components, and no additional features and functions could have been programmed later. A processor-based product would have used software to emulate the functions of a 029, or—since it was a computer—anything else the programmer set out to do.)

"I said wait a minute, I realize that it is much simpler to hard-wire a device, but no one here really knows what it will be used for. The vision is to use it for all sorts of things—word processing, business management, general data processing, even running a nuclear reactor. If you're going to make this a single-purpose machine you are doing it on the assumption that the 029 replacement market even exists, and we have not tested that yet. I recommend you stay with the original plan for a general-purpose processor," Frassanito recalled saying.

Things got heated. "I said, 'God damn it, you will kill it, you are rolling the dice, you're narrowing its applications down to one that might not exist. If you are going to go out of business I might as well yell at you now.' I would not back down. I went back to body shop mode, where you yell a lot," he said.

He made no real progress. The engineers finally conceded that they might get around to it, eventually, but that they were very busy with "sustaining engineering" for the Datapoint 3300. It was clear that they really didn't want to do it.

"Phil and Gus came back the next day and said they had thought about what I said. They said they had been planning to do a hard-wired version first and then do a processor version. I said that it takes so long to do a new product that you had better do the right one the first time. You might not sell any of the hard-wired versions and then it would be another year before you do the processor version.

"'Frazz,' they said, 'You were pretty hot in there, but we think that is what we will do.'"

And so the decision was made—but it had also become clear that the new product would require a new design team.

Project Brinksmanship

Later, CTC's management set off on another round of fund-raising to finance the development of the new product, dubbed the Datapoint 2200 since, with its half-height screen, it would be smaller than the 3300. Frassanito produced a drawing of a unit with an integrated keyboard like the 3300, but with a cassette tape drive built into the top. Between the bottom of the screen and the top of the keyboard was a row of flat toggle switches, similar to the ones found on the face of the DEC PDP8, used for inputting code during boot-up. If the device was going to be a computer, backers would expect to see those switches, he reasoned. (The final version dispensed with the toggle switches, but added a second tape drive to the top.)

Ray, Roche, and Frassanito proceeded to develop the business model for the machine, write the business and financial plan for it, illustrate the concept with a polished product picture that included the toggle switches, and make pitches to investors. The plan included the assumption that CTC would supply canned software for specific applications.

The money men accepted the premise, and the picture. But CTC wanted $15 million to develop the machine, and they were only able to raise $5.6 million, Frassanito recalled.

They decided to make do.

They also made a conscious decision to stick with the methods that they had learned in the space program, an approach that NASA insiders called "buck whipping" and which others nervously described as "pouring money on it."

No, they didn't have all the money they wanted—but if they showed progress, they should be able to raise more money later. The thing they could not raise more of was time. So at any point where there was a choice between saving money or saving time, they would choose to save time. By finishing the project faster, they should save on overhead in the long run anyway. The trick was to peg the deadline to an event outside the organization, such as an upcoming trade show (as they did with the 3300) or a martyred president's pledge to get the country to the moon by the end of the decade (as they did in the space program.) That way the deadline was not subject to renegotiation.

They had seen the approach work in the space program and with the 3300, and they were certain it would work for the 2200, Frassanito recalled. But to their conventional investors and business associates, the results looked like one long spending spree.

With the basic decisions made, and the company committed to the project, the time had come to find a team of engineers who could and would create the machine.

The Thanksgiving Instruction Set

At this stage, Gus called on his old friend, Vic Poor, who had coached him on what to say when talking to potential backers.

As previously mentioned, at the time Poor was the former principal of a custom electronics design and manufacturing firm in Frederick, Maryland, called Frederick Electronics. It had been bought out by a budding conglomerate, Pacific Plantronics. Poor could see that the new owners would not need him, and so he was looking for new opportunities.

"Primarily we made high-frequency radio hardware, such as receivers and demodulators," he recalled about Frederick. "We also did telex switching systems, including store-and-forward systems. And we did telegraphy systems to send and receive Morse code—the Navy was a big user."

Poor himself was a big user of telegraphy systems, being an avid ham radio operator. As for those store-and-forward systems, they had involved routing telex messages to the right circuit based on header information in the messages themselves. Poor recalled that they had kicked around the idea of designing their own basic computer to handle the task, but had not found the time to do it. That was about to change.

The call from San Antonio came after the end of summer in 1969. "Gus asked if I would come to work for him, as their technical director. I visited and decided it was worth a fling," Poor said.

Oddly enough, before he could start his new job at CTC he then came down with an appendicitis, and was recuperating from the appendectomy at his house in Frederick, during the 1969 Thanksgiving holiday, thinking about the processor for Roche's machine. There, he was visited by two employees, Jonathan Schmidt and Harry Pyle. The latter was actually a part-time employee and a student at Case Western Reserve University in Cleveland, where classes were out for the holiday.

Pyle was then about 20. He had gotten to know Poor through ham radio activities while in high school.

"I lived in Wilmington, Delaware, and he was not far away (in Maryland) and had good (ham radio) signals. Hams had figured out how to use old Baudot[32] Teletypes. His (Poor's) mother was part of a missionary service that used jungle aviation to deliver Bibles. They were using ham to send Teletype messages, and they needed some piece of gear, which I built with tubes. Poor was impressed enough to offer me a summer job," Pyle recalled four decades later.[33]

He became familiar with configuring communications controllers with 16-bit computers

---

32 Baudot was a previous-generation five-bit Teletype code that supported fewer characters than the 7-bit ASCII code. ASCII had become standard on new hardware by 1969.
33 Quotes are from an interview with Pyle in December 2007.

from Interdata Inc. of New Jersey. "I started programming it at night. That's when Poor started to appreciate my abilities," Pyle said.

Indeed, Poor liked to tell the story of returning from a business trip and dropping by the office before returning home—and finding that the firm's computer was playing Anchors Away. Pyle was sitting there feeding punched paper tape into the tape reader of the Teletype that was used as the computer terminal. Pyle had found that the computer put out a lot of radio frequency noise, but that the noise varied according to the instruction the computer was performing. He had mapped its operations to a musical scale and was feeding it code to run appropriately timed loops of instructions. He sat a radio on top of the computer and tuned it to pick up the radio noise, which, thanks to the instruction loops, sounded like music.[34]

Then 28 years old, Schmidt had received his ham radio license at age 12, and had graduated from high school at 17. Four decades later he still displayed the unruffled demeanor of someone who had spent the third grade in an iron lung as a result of polio. It left him unable to raise his arms over his head without throwing them, but had not stopped him from running 25 marathons.

His ham activities had led him to spending a year and a half on the SS Hope charity hospital ship. He remembered piping news bulletins about the assassination of President Kennedy into the ship's public address system as it transited the Panama Canal.

One day, off the coast of Ecuador, he picked up a surprisingly strong signal from another ham back in the U.S.

"He said he was in a travel trailer with a borrowed Collins radio," Schmidt recalled. "He gave me the model number and said he borrowed it from a guy named Irv in Ann Arbor. I said that I was Irv's roommate in college and that radio was supposed to be stored in his closet since it was mine."

The ham he was talking to was Vic Poor. "We became fast friends. It turns out I had designed the demodulators used on the ship with a device patented by Vic Poor at his previous company. After I went home I got a master's in Ann Arbor and had done the coursework for a PhD, but Poor kept increasing the offer for me to quit and work for him, which I did in 1966," Schmidt explained.

At one point they were debugging a radio demodulator, built by a British firm, which was used to receive pictures transmitted by the Associated Press. The pictures came through fine, but each picture was marred by the presence of an offset shadow of that picture. Eventually Schmidt concluded that there was nothing wrong with it except that it was too well designed. It picked up the transmission of the picture when it first arrived—and then when it arrived a second time after circling the globe. They added a feature to mask the low-level signal—make it less sensitive, in other words—and the customer was happy.

That 1969 Thanksgiving, Schmidt recalled watching Pyle and Poor design the PC processor.

"We did it on the floor. The cat was hanging around," recalled Pyle. "Vic had done some

---

34 Today, the U.S. Government will not allow the sale of a computer that emits that much radio noise.

thinking. I was interested in computer architecture design, and saw it as an exercise in minimalism. Most machines then were 16 bit, and Vic had the idea of taking two general eight-bit registers and using them for the memory address," Pyle recalled. "The instruction set had room for eight registers, A through E, plus H and L for the high and low half of 16-bit memory addresses. And then there was the M register for the memory location that H and L were pointing to."

"One of the few things that I insisted on was a subroutine call stack," Pyle added. "Vic wanted to jump back (after a subroutine) to one of eight locations based on the conditions of a set of bits, but I said that we needed a subroutine calling mechanism, and set up a dedicated subroutine call stack."

Soul of the Machine

The instruction set defines what a computer can do, and the instructions are the basic tools of the programmer. Each instruction represents an extremely basic task, usually centered around fetching a byte from a memory location, doing something with it if the byte represents data, or executing it if the byte represents a command. Applying the instructions to perform a real-world task is like solving an intricate puzzle. Using them directly is called machine language programming, and is rarely done anymore. Modern programmers use higher-level programming languages that can define a task in something resembling human-language syntax, and are thereby insulated from the arcane intricacies of machine language. But those higher-level languages—and ultimately all other software—run in machine language, because that is all that the computer "knows" how to do.

Then and now, anyone doing significant amounts of machine language programming would use an intermediate level of software called an assembler. An assembler, among other things, lets the programmer input the machine-language instructions as three-letter code names and invoke subroutines by name instead of having to track their locations in memory. An assembler would be one of the first pieces of software that the CTC staff wrote for the new machine.

The instruction set they created during that holiday covered the minimum number of operations needed to compute: their planned machine could load data from memory into a register and there perform addition or subtraction or various binary operations (AND, OR, XOR) plus compare, and shift right or left. It could also jump to a location, call a subroutine, return from a subroutine, input data from the outside, halt, and (for timing purposes) perform "no operation." Counting variations of these themes, there were 31 instructions. A modern x86 processor has about 400, but equivalents of all the original 31 are still there.

A modern PC uses Random Access Memory (RAM) for its internal memory, and the processor can directly access the contents of any available memory address, in any order, at any time (hence the "random" name.) For the Datapoint 2200, RAM chips were not available, and the hand-wired magnetic cores used to make RAM for mainframes would have been far too large and expensive. (For instance, 4K of core RAM for the DEC PDP-8 costs $10,000 and had to be mounted in a rack. Each bit was embodied in a tiny magnetic donut strung on thin wires.) So for memory they planned to use shift registers in blocks of 2,048 bytes. The range of addresses rotated past the virtual read-head every 520 microseconds. (A microsecond is one-millionth of a second.) Therefore, a specific memory address would go by about 2,000 times per second.

Consequently the memory would be much slower than the speed of the processor, which would run at 125,000 cycles per second.[35] The processor took at least two cycles to do anything, meaning it could perform at most 62,500 instructions per second.[36] That made it more than 30 times faster than the memory. Consequently, any time the processor needed to access a memory address, it faced a dramatic slowdown.

To get around the problem they used registers for performing nearly all tasks. A register is a scratch memory cell inside the processor into which the contents of a memory address would be loaded. All operations would be performed on the data while it was in the register, and the memory access delay would only have to be faced at the start of a process (and possibly at the end also, if the results had to be placed back in memory.)

Another trick was based on that fact that the rotation of the memory addresses looping inside the shift register could be stopped and restarted at will, since there was no momentum. Therefore, the program could always step from one memory cell to the next without incurring additional delays. Pyle recalled spending a lot of effort optimizing the software for the Datapoint 2200, so that necessary addresses were arranged sequentially. Therefore, their contents could be retrieved without additional delay.

The use of sequential addresses became unnecessary after RAM chips became widely available, about two years later. But the reliance on registers remains a hallmark of the x86 architecture to this day. To uninitiated machine-language programmers the approach looks unnecessarily clumsy. Surely it would be simpler to skip the "load the contents of the memory cell into the register" step and just perform the operation directly on the contents of a memory cell. Later, some non-x86 processors did just that. However, it was found that such processors did not work well with software stored in ROM, since ROM memory cells could not be changed. Decades later, the situation faced by the CTC designers returned—in modern PCs, to prevent over-heating of the dense memory components, the RAM is many times slower than the processor. So the decisions that Poor, Pyle, and Schmidt made over turkey remain applicable.

Beyond those considerations the Datapoint 2200 was slow, even by 1969 standards. For instance, the era's iconic small computer, the previously mentioned PDP-8, had a processor clock frequency of about 830,000 cycles per second. It not only had RAM instead of shift registers, but its RAM was nearly as fast as the processor.

Another reason for the 2200's lack of speed was that the machine looked like an 8-bit computer but was actually a one-bit or "bit serial" computer, just as Ray discussed with Bender in 1969. (See Chapter 3.) The previously described machine instructions used by the programmers were 8-bit operations, processing full bytes. Internally, calculations were

---

35 The top speed of modern PCs is about 4 gigahertz (four billion cycles per second) or about 32,000 times faster than the Datapoint 2200. At this writing, running much faster than 4 gigahertz is impractical because of heat dissipation problems, and the vendors are offering higher power through multiple cores rather than higher clock speeds.
36 Modern high-end PCs are rated at hundreds of millions of operations per second.

handled one bit at a time, starting with the right-hand or lowest-value bit of the byte in question, and then cycling through to the left, to the other seven bits, in turn.[37]

(Starting from the right is called little-endian, and all PC processors since had inherited this little-endian internal approach, to the confusion of generations of machine-language programmers, since natural language is typically big-endian. For instance, 84 is typically spoken big-endian style as "eighty four," rather than little-endian "four and eighty." IBM mainframes were big-endian. The terms were borrowed from "Gulliver's Travels," which depicts a bitter war between little-endian and big-endian factions, although in that case they were referring to which end of the egg they broke to serve breakfast.)

Poor said he adopted the bit-serial route because he wanted to minimize the number of components that would be needed to make the processor—because he expected to have it done on one chip.[38] If in the process the Datapoint 2200 was slower than it might have been, that wasn't an issue because it was just a "programmable terminal" and didn't need to be fast, he said.

That brings us to two points of controversy. The first concerns the nature of the machine they thought they were designing. The second is bigger.

"Programmable Terminal"

Basically, Poor, Schmidt, and Pyle agree that they were designing a programmable terminal, a machine whose main attribute was that it could be programmed to emulate more computer terminals than just the Model 33 Teletype. This would broaden the market that CTC could appeal to. Later, the programmable terminal was found to make a good desktop computer.

That was the story the author heard when taking sales training at Datapoint Corp.[39] in 1980. The story surfaced almost three decades later, when the author wrote a story for Computerworld[40] outlining the origins of the 2200 as a purpose-designed desktop computer. Several readers posted comments complaining that the story was wrong, citing the old story that the 2200 was a programmable terminal.

Some versions of the story say that CTC's management was surprised when the buyers began programming their 2200s to perform end-user applications. CTC's startled managers then re-envisioned the 2200 as a desktop computer and began marketing it as such. The implication is that the invention of the personal computer was an accident.

The idea that the personal computer was created by accident doubtless appeals to those who like to think of mankind muddling through, and of things turning out right despite everything. But the evidence shows that there was no such accident.

The evidence is that Roche (and probably Ray as well) intended from the beginning to make

---

37 According to Michael Fischer, serial designs were rare but hardly unknown at the time. For instance, there was a serial version of the previously mentioned PDP-8. Serial designs could be found in low-end microcontrollers into the mid 1990s.
38 As noted in Chapter 3, Phil Ray had been talking about a one-bit computer in May 1969.
39 As will be discussed later, CTC was renamed Datapoint Corp. in late 1972.
40 The story ran August 8, 2008, at www.computerworld.com.

a desktop computer. The business plans, Ray's last interview, Frassanito's recollections of his conversations with Roche months before development started on the 2200, Benders interview with Ray in Boston in 1969, the magazine ad made for CTC in 1970 (discussed in Chapter 7), and Roche's July 1971 interview as mentioned in Chapter 1, all indicate that the 2200 was intended from the start to be a computer.

Beyond that, the CTC annual report for 1970 touted the 2200 as a general purpose computer. "Through programming of the self-contained control computer, the Datapoint 2200 may be used for an infinite variety of data processing applications," it said.

There's also Ray's message, as president of CTC, in the forward of the 1971 annual report, flatly stating that the 2200 was the founder's original goal when they started the firm. He wrote, "In 1968, when Computer Terminal Corporation, your company, was formed, its principal product objective was to develop and successfully bring to market, at a reasonable price, a truly flexible intelligent source data capture terminal in the shortest possible time frame. During Fiscal 1971, shipments of this product, the Datapoint 2200 Intelligent Terminal, began."

As will be shown in Chapter 7, the first end-user customer of record bought the Datapoint 2200 specifically to use it as a desktop computer, the CTC salesman having represented it as such. (Actually, the 2200 would seem like a poor choice for a terminal emulator, with its half-height screen. Also, its base price was about double that of the 3300.)

Chris Roche, in fact, recalled that during this period Frassanito presented Chris' father (Gus Roche) with storyboards for a proposed Super Bowl commercial touting the 2200 as a computer for home use. The elder Roche nixed the idea, saying the 2200 would be a business product priced above the home market. (Frassanito remembered it as an effort to increase public awareness of computers.)

The original programmable-terminal story, incidentally, turns out to be more complicated than the "accidental computer" legend implies. "The idea was to make a programmable terminal that could emulate a terminal from Burroughs or CDC or any other computer vendor," Poor recalled. "Those companies had terminals specialized for their computers and protocols, and we would write a program for our machine so we could sell to those markets. Did we aim at the 029? Yes and no—we wanted something that could do electronic data entry, capture the data on cassette and transmit it over a serial (i.e., telephone) line. We were not terribly focused at that point, but the focus would come from the programming."

It other words, it was supposed to be a general-purpose device—a computer. It sat on a desktop and was intended for use by one person, like a terminal, so comparisons with a terminal were inevitable. Moreover, programming it to emulate specific brands of terminals was an obvious marketing angle.

In any event, the controversy tends to fade when you examine it closely. A programmable terminal, or a general-purpose device that could be programmed to emulate a terminal, is a computer—otherwise it would not be programmable. At the time, calling it a computer would have implied that it was a massively expensive, room-sized assembly of hardware—or at least a processor with a significant amount of memory. Originally memory was so expensive that the base amount on the 2200 was only 2,000 bytes. That is barely enough memory space for a self-respecting machine-language tic-tac-toe program. But if the 2200 were to be sold as a computer, the customers would expect a programming language and an operating system, and 2K was not enough to support that.[41]

Hand-coded terminal emulation programs could, however, have been shoe horned into such a space. The machine could be sold with software for specific terminal emulations (and other applications) and used as an appliance, with no more consideration to end-user programming than is given with the microprocessors in modern microwave ovens. Additionally, CTC knew that there was a certain amount of demand for terminal emulators, while the market for desktop computers had yet to be born. Indeed, no one could be expected to know what desktop computers were.

But (as shown in Chapter 7) it turns out that CTC immediately ran into customers who understood exactly what the 2200 was and were willing to pay for the memory needed to use it as a desktop computer. CTC then scrambled to meet their expectations.

Incidentally, the cost of memory did not fall to a penny per bit (not byte) until 1972, at which time filling the entire 8K memory capacity of the 2200 would have represented a component cost of "only" $640.

The 54 page CTC software catalog for October 1972 (the earliest available) had only two pages devoted to terminal emulators. The seven packages on those two pages included one that would run in 2K, to emulate an IBM 2741 hard-copy terminal. It was one of the few packages in the entire catalog that would run in 2K. There was also an emulator for the IBM 2780 combination card reader and line printer, which ran in 4K. Almost everything else required 8K. (There was no listing, incidentally, of an IBM 029 card punch machine emulator, implying a lack of customer interest by then. Apparently, Frassanito had been right to worry about tying the fate of the 2200 solely to the 029 emulation market.)

The other, bigger controversy concerns who really invented the microprocessor chip. If Vic Poor intended the processor to be reduced to a single chip, and then actually got it reduced to a single chip, does that make him the inventor? Or is it unfair to say the device had a single inventor?

The issue is complex—as the next chapter shows.

---

41 The original operating system for the Datapoint 2200, based on cassette tape and called CTOS, needed 6K.

Chapter 5

Sparking the Microchip: Late 1969, Early 1970

Vic Poor began working for CTC in San Antonio in December 1969. One of the first things he discovered was that Roche and Ray wanted as much of the 2200's processor circuitry as possible to be one chip. They wanted the 2200 to be, basically, the same size as an IBM Selectric typewriter, that being the piece of equipment in widest use in American offices. But to get all the anticipated circuitry into such a small box would take some crunching. Maybe Intel could help.

Almost immediately he paid a visit to Intel about various matters and also to inquire about a possible custom processor chip. CTC was Intel's biggest single customer for shift registers, as a result of the fact that the devices were used to store the contents of the screen in the Datapoint 3300. In fact, some sources say that, at the time, CTC was Intel's biggest customer, period. Consequently, Poor had their full attention.

In a 2004 interview,[42] Poor recalled talking to five different people while he was at Intel, and hearing that Intel was not interested in making small quantities of exotic, custom chips. They were only interested in large, commodity orders. They offered him a four-bit calculator chip they were developing (the future 4004), which Poor rightly rejected as unsuitable for the task.

Apparently, the last person he met was chip designer Stan Mazor—who said that the chip that Poor wanted could be made.

"We would have liked to have an MSI part specific to our requirements to implement the instruction set. I talked to Stan Mazor out there about what we needed. He said they could put all this on one chip and I said that was fine, too," Poor told the author in a 2008 interview.

Mazor[43] himself gave a much more detailed account of that meeting, which he remembered as being around Christmas 1969. Like quite a few other early Intel figures, Mazor had previously worked at Fairchild Semiconductor, starting in 1964 as a programmer, becoming a computer designer in 1966. He had joined Intel three months earlier.

Poor's visit that day was mainly to check on CTC's shift registers, which were being custom-made by Intel—in fact, Intel had a special part number just for the CTC shift registers, Mazor recalled. Poor originally met with Intel business managers, including Andy Grove.[44] Mazor recalled that at the end of the meeting Poor expressed interest in Intel's new 64-bit RAM circuit, which he felt could be turned into a stack pointer with the addition of an address router.

"Grove is a brilliant physicist but he knows nothing about logic or chips, so he calls me into

---

42 The 2004 interview was recorded by the Computer History Museum. Poor reiterated most of its points to the author later.
43 Quotes from Mazor are from a phone interview in 2009 followed by a face-to-face meeting at a computing conference. Mazor also supplied background information.
44 In the 2004 interview, Poor recalled meeting Gordon Moore, Bob Noyce, "the Hungarian" (Andy Grove, born Andras Grof), Ted Hoff, and Stan Mazor.

the meeting," Mazor explained. "Alone with me, after Grove leaves, Poor said he needs a 512-bit push-down stack[45] chip, and could I make this chip into a stack. It was a 15 minute discussion. I asked what the stack was for, and he said he was making an eight-bit computer and wanted to save some of the registers in the stack. So I pulled out a tablet and made three proposals.

"The first one was eight registers—64 bits—with a push-down stack on the same chip. But I knew that we could do better.

Then I tore off the sheet of paper and suggested that we do that, and also add an 8-bit arithmetic unit, to do arithmetic with the registers," Mazor continued.

"With the third sheet of paper I said that it may be the case that we can do the entire CPU on one chip. Since September, three months earlier, I have been working with Ted Hoff on a four-bit CPU which had 15 four-bit registers. It was not a big leap to make an entire eight-bit CPU on a chip."

"What I said to Poor (suggesting a one-chip CPU) was unbelievable. It's like a guy asking for a one-ounce container and I suggest a two-gallon pail—it was 20 times more than he was thinking of. But we were already working on a four-bit CPU."

"I said that whether it was possible depended on how complicated the computer was, and he agreed to send me the programming manual that described the CPU. That was the end of the conversation."

"Then I went to my boss Ted Hoff and we did an outline, Poor sent us the manual, and we sent them a proposal for the chip. I deleted several instructions (from the Datapoint 2200 instruction set) that I thought were too complicated, like branch-on-bit," he recalled.

Poor recalled returning to San Antonio and telephoning Pyle, who had returned to college, to write a one-chip specification for the processor. They called it a "chip processor."[46]

The Predecessor

The four-bit processor project (later named the 4004) referred to above was a troubled but ultimately successful effort to develop a processor chip that would run a scientific calculator. Historians often refer to the 4004 as the first microprocessor, and imply that the eight-bit processors that came afterward were extensions of the 4004. Actually, it was a stand-alone project whose final product was incompatible with the CTC project, or anything else that would later become the x86 line.

Its data path was only four bits wide—sufficient for handling numbers, but inefficient for handling text, as computers do. But that was only the least of the crippling limitations that prevented it from being used as a general-purpose computer, and no one ever used it in one. It could not write to program memory, or access or jump to program memory beyond the 256-byte page of memory that was currently loaded. There were only three

---

45 A push-down stack is often compared to the stack of dishes in a spring-loaded well near the front of a cafeteria serving line. A new plate goes on top and pushes the stack down. The next plate used is removed from the top, in last-in, first-out fashion. Data bytes are handled similarly in a push-down stack.

46 "Chip processor" is from an interview with Poor in the January 1976 issue of Datamation Magazine.

basic data manipulation functions. It had no Boolean operators, which programmers use to directly compare the contents of memory cells. As a result, it could not have run an operating system. As a processor, it was so inefficient that adding two eight-bit numbers (a fundamental task) required 79 instructions.[47] Beyond that, it required a custom memory chip, incompatible with any other system. Conversely, standard computer memory would not work with the 4004. But these limitations were immaterial for the chip's intended use in a desktop calculator, where it merely had to process individual numbers fast enough to keep up with the operator's manual input. But it was clearly not in the same league as the chip that the Poor-Mazor meeting would lead to. That chip would embody the functions of a general-purpose computer, and it did run an operating system.

Meanwhile, as this book will show, the fact that it appeared on the market before the Datapoint 2200 chip was purely accidental.

The 4004 project[48] had gotten under way the previous June when executives from a Japanese consortium called Busicom paid a visit to Intel to talk about producing a scientific calculator. The group included an engineer named Masatoshi Shima. They chose Intel because of its reputation as a pioneer in the silicon-gate chip-making process, which allowed for more densely packed chips than did the previous-generation metal-gate process.

A scientific calculator, of course, would need to perform transcendental functions, such as trigonometry and exponents, requiring much more logic than the arithmetic functions (i.e., add, subtract, multiply, divide) expected on a consumer calculator. Busicom presented a complex logic layout that required 12 chips. Intel countered with a new approach that only required four chips. One of the chips would be a tiny, basic computer—the processor. Later called the 4004, it would perform math operations based on software. The other chips would be the 4001, which would store the machine's program in ROM;[49] the 4002 with internal scratch memory in RAM; and the 4003, which handled I/O for the 4004 using a shift register.

The Japanese executives agreed. Shima made another visit to Intel in September 1969 to settle certain details. He planned to return in April 1970 to check the final design.

He would be sorely disappointed.

Fixing the Base

Back at CTC, sometime in early 1970 Poor and Mazur followed up on their late December meeting and agreed that an eight-bit microcomputer chip that embodied the Datapoint 2200 processor could be developed. But that did not mean there had been any decision to launch any such development project—that was a business decision, not a technology issue.

Here the trail grows momentarily cold because no documentation of the business deal between Intel and CTC concerning the Datapoint 2200 chip has survived. The Intel archives

---

47   Personal communication with Michael Fischer.
48   Additional background information about the 4004 project are derived from a series of history interviews and first-person articles that appeared in Electronics Weekly in 2008.
49   ROM stands for Read Only Memory, and can be thought of as RAM whose contents are set at the factory and cannot be changed by the user. On the other hand, the contents of ROM survive when the unit's power is turned off.

have a copy of the development deal concerning the 4004, but, curiously, nothing about the CTC deal. All the sources on the CTC side of the story agree that Intel had to be "dragged kicking and screaming" into the deal—a categorization that the people on the Intel side of the story have long resented.

Frassanito recalled first hearing about the possibility of a CPU chip after getting into an argument with Poor about the size of the enclosure of the Datapoint 2200—the design engineers could not quite get the processor board to fit. Words were exchanged, including choice New York vernacular, he recalled.

"If you have a problem, we'll fix it—but you get on my case with every technical issue that comes along and I'll punch your lights out," Frassanito recalled yelling before storming out and slamming the door so hard it came off its hinges, cracking plaster in the adjoining wall.

Later that day, CTC president Phil Ray called Poor and Frassanito into the conference room to make sure the issue was settled and that peace prevailed. The room contained a preproduction prototype of the Datapoint 2200. There were periodic meetings in that room to discuss the features of the 2200 and various engineering problems. There was constant emphasis on making everything inside it as small as possible, to save space and reduce the heat that the unit generated. (One of these meetings had resulted in the removal of the row of flat toggle switches above the keyboard, which had been inspired by the DEC PDP-8. The 2200 could automatically load software from a tape drive, so the switches were not needed.)

The morning's problem, they decided, boiled down to the fact that they had already committed to the tooling of the enclosure. In other words, the expensive steps that had to be taken to mass-produce the case of the Datapoint 2200 had already been taken, and there was no way to change the design without crippling expense and delay.

The base plate (the bottom of the unit) was to be manufactured by a simpler, less complicated sand-casting process. Therefore, changes could still be made to the design of the base plate without wrecking the production schedule and budget.

Frassanito suggested indenting the base plate by an eighth of an inch, leaving room for the processor board.

The team agreed to do that, and the problem was solved—production of the Datapoint 2200 could proceed with a board-sized processor. The bottoms of 2200s are marked with a one-eighth-inch bump as a result.

Ray, however, wasn't satisfied, Frassanito recalled. He took one of the chips that were there for the prototype and used an X-Acto knife to pry off the top, since it was made in two parts. As science was his hobby, he had a microscope in his office, of the type used in schools. He went to his office and returned with his microscope, and put the chip under it. He examined the circuits, and he announced that, as far as he could see, there was no reason that they couldn't get all the components of a processor board onto one chip.

Gus Roche was especially intrigued with the idea. Yet, having gotten the processor board to fit inside the Datapoint 2200, they did not actually need the chip. Had someone agreed to make the chip, they would not have dared change the 2200's production plans until deliveries of the chip were firmly in hand. But Roche apparently saw the chip as a path toward the eventual expansion of the 2200, since it would open up space inside the unit.[50]

"Gus and I felt strongly that what the world needs is a computer on a chip, just to make it smaller and use less power," Ray recalled in his last interview.

So, at some point in early 1970, Frassanito found himself accompanying Roche to Intel headquarters in California to meet with Intel president Dr. Bob Noyce[51] to, basically, negotiate the development of the world's first microprocessor. He remembered being ready for anything—except for what actually happened.

Keep in mind that Intel was another start-up like CTC. CTC, as mentioned, was a major customer of Intel, and the relationship had warmed to the point where Roche had given Noyce a Datapoint 3300 for his children. So they had no difficulty setting up a meeting.

Pushback from Intel

Frassanito would later remember a host of small details, such as flying from Texas the previous night, staying in the Hyatt Hotel in San Francisco, and driving out to the Intel offices in Silicon Valley through a morning fog. They were immediately ushered into Noyce's office, which he remembered as being unpretentious, with steel and Formica furniture. It was bereft of ego toys. Instead of a power suit, Noyce was dressed in gray slacks and a tweed jacket. He struck Frassanito as a friendly, erudite gentleman.

The meeting went on for more than three hours. Dr. Noyce sat by himself, behind his desk. No one joined him or interrupted him, and his phone never rang. But while they had Noyce's attention, he seemed on edge, Frassanito recalled. Perhaps he was already wrestling with reservations about the subject.

The proposition that Roche presented was simple: Intel could take the instruction set of the Datapoint 2200 and—free of charge to CTC—develop a processor chip that emulated it. Intel could sell the chip to anyone willing to buy it. All CTC asked was that CTC could buy it, too.

He remembered Roche saying that a low-power microprocessor chip, manufactured for a few cents, would be a compelling, important, and elegant product. Computers could be churned out in vast quantities, and the market would surely lap them up. The world surely needed such a chip—certainly CTC did, as it would help reduce the heat generated by the Datapoint 2200. (Noisy fans were the only alternative.)

They then turned to technical requirements. Roche was by then aware of the four-bit 4004 chip that Intel was developing for a Japanese firm. It would need about 2,300 transistors. Emulating the eight-bit Datapoint 2200 processor (which sat mostly on one board with about 120 chips) would require somewhat more transistors, but it was assumed they could do it.

(The fact that they could count on being able to put additional components on a chip in the

---

50 Poor, in the 2004 interview, confirmed that the chip was seen as a future replacement for the processor board.
51 Noyce was a co-founder of Intel and was its CEO until 1975. He died in 1990.

near future was called Moore's Law, named after another Intel pioneer, Gordon Moore, who did not attend the meeting. At the time, it implied that the number of devices—transistors, resistors, and diodes—that could be economically put on a chip would double every year. The doubling rate has since slowed to about 24 months. At this writing the transistor count exceeds two billion per chip. Fulfilling Moore's Law involves achieving steady incremental improvements to the planar system of making chip circuits. Noyce had been involved in developing the process at his previous job at Fairchild Semiconductor.)

Frassanito remembered Noyce seemed to be vacillating on whether he liked the proposal. Finally, when the arguments had all been made, Noyce stood up, put his hands on his desk—and turned down the idea.

"It's intriguing, and we can do it," Frassanito remembered him saying. "We can take a shot at it—but it's a useless product, and making it would not be in our business interest."

Noyce then went on to say that he did not see much market for such a thing. He added that Intel was in the memory business. Whenever someone made a computer, Intel could sell that vendor hundreds of memory chips. If Intel made processor chips, Intel would only be able to sell one chip per computer, Frassanito recalled.

In hindsight, those seem like weak, arbitrary arguments. Actually, at the time they were quite compelling, so much so that it's a wonder that CTC ever overcame them.

As for Noyce not being able to see any way to market such a thing—well, there was no market for microprocessors since no microprocessors then existed. Of course, marketing products that had not previously existed was what the electronics industry was all about—but in this case it must have been hard to envision a market big enough to justify the effort. After all, there were fewer than 200,000 computers in use worldwide at that time. The previously mentioned DEC PDP-8 was considered a very successful small computer, but the production run of the original model, sold from 1965 to 1968, was only 1,400 units.[52] (It would sell 50,000 by the time it was discontinued in 1990.) The whole minicomputer market had seen sales of only 6,000 units in 1969.[53] By the end of 1972, CTC (then renamed Datapoint) would be happy to brag of selling 160 2200s per month.

A calculator chip like the 4004 was another thing entirely, since calculators could be turned into mass-market items with high sales volumes. Japanese shipments were expected to reach 2.75 million in 1971.[54]

Meanwhile, figuring out how to make such a chip would just have been the start of the problem, since sales and support for processor chips would be radically different from the sales and support of memory chips. After all, memory chips were used about the same way in every computer, but with processor chips, every design would be different, and the customers would need vastly more engineering and programming help.

As for Noyce's comment about only selling one chip per computer, he must have meant that, if Intel started selling computer chips, the computer makers that bought its memory chips

---

52 See http://www.faqs.org/faqs/dec-faq/pdp8-models/section-2.html, accessed July 3, 2009.
53 The IEEE Computer Group News, July-August 1970, citing a report by Arthur D. Little Inc.
54 Electronics, November 23, 1970, page 84.

would see Intel as a competitor, and buy memory chips elsewhere. Intel's memory business would go away and it would be left with only its processor chip business, selling a mere one chip per computer.

Roche and Frassanito did try to say that Intel could just as easily sell both memory and processor chips. Predictably, Noyce wasn't swayed. They did not try to argue that processor chips could be sold for a lot more money than memory chips, having just undercut that argument by stressing how cheap microprocessor chips could be.

Frassanito could see that Roche was passionate about the idea of a microprocessor chip. Roche pressed ahead with his final argument, offering to pay for the chip's development. Presumably, he may have foreseen that just asking politely might not suffice. He suggested (as Frassanito recalled) $50,000 with a six-month deadline, with CTC keeping the intellectual property.

Frassanito got the impression that Noyce was agreeable, and the meeting broke up. Outside they were introduced to other Intel executives, some of whom, like Moore, later became famous.[55]

Conflicting Stories

But apparently Noyce's decision in that meeting was not the last word. Literally every source has a different interpretation of what happened next.

Frassanito, as noted, felt that Intel had agreed to make the chip for $50,000. Such an agreement would have been similar to Intel's agreement concerning the development of the 4004 chip, as will be explained shortly.

Mazor remembered that CTC agreed to buy 100,000 units for $30 each, but felt that such a huge quantity would only have been realistic if they were selling memory chips. "It was ridiculously too many units—it showed that the commercial side of the operation did not know what they were doing," he said.

Ted Hoff, Mazor's boss at Intel, remembered that the contract amount was 30,000 units—a lower but still an unrealistic quantity.

Intel chip designer Federico Faggin remembered the contract deadline as being nine months, starting in March 1970, but did not remember the development fee—except that it must have been "tens of thousands."

Vic Poor at CTC remembered that the agreement with Intel involved non-recurring expenses of $100,000. He assumed that Ray got Intel to agree by threatening to take CTC's chip business to Texas Instruments

But Phil Ray, in his last interview, insisted that the chip was developed on a $100,000 bet. Intel would get $100,000 if it could deliver the chip in six months, and would have to pay

---

55 For Frassanito at least, it would be ironic that Moore would get credit for the law that governed the progress of the microprocessor industry, yet his company had initially not wanted to make microprocessors. Fischer, however, remembers Moore being personally in favor of the idea of making such devices, despite the microprocessor chip marketing being, at first, insignificant.

CTC $100,000 if it could not. Supposedly, the bet was sealed at a dinner party. (It is not clear from his description if Ray was betting with Noyce himself, or a subordinate.)

Unlikely as that sounds, keep in mind that Intel did not go public until the next year, so its management was not at that time shackled with the kind of fiduciary conservatism that marks most investor-owned corporations. (CTC was already a public corporation, but winning the bet would save it money. Losing meant getting what it wanted faster.)

Interestingly, page six of Intel's in-house 16-year anniversary celebration brochure ("A Revolution in Progress," published in 1984) speaks of a "company tradition of betting as a management motivational tool." However, the stakes were usually bottles of champagne, not tens of thousands of dollars of the company's money.

Meanwhile, Bob McDowell, Intel's sales representative in Texas at the time, said he was not privy to the negotiations behind the development of the chip, but remembered that the purchase order for the project was fairly simple. However, he remembered that it included a provision that the principals referred to as a side bet, but which the document referred to as a penalty clause.

"It said that if either party canceled there would be a $50,000 penalty against the party that canceled the agreement," McDowell said. But the $50,000 penalty was over and above the basic development fee, and McDowell could not remember that amount. Aware of its significance, he kept his copy of the purchase order for years, but at some point misplaced it.

A possible reconciliation is to assume that Frassanito was correct about the development fee being $50,000. Adding the $50,000 penalty could supply a $100,000 total amount for the bet that Ray spoke of, and the $100,000 that Poor remembered.

The 4004 Contract

It seems logical to assume that the 4004 contract was used as a precedent for the CTC agreement. In any event, its provisions must have been negotiated by the same people at Intel who were involved in the CTC contract, during roughly the same period. Presumably, they were written along the same lines.

Nine pages long, it was signed on February 6, 1970, between Intel and Nippon Calculating Machine Corp. of Osaka and ElectroTechnical Industries Corp. of Tokyo. The development fee was $60,000—more than CTC was apparently charged, but on the other hand multiple chips were involved. The contract described the four chips that Intel was to develop (i.e., the 4001, 4002, 4003, and 4004, although the chips had not yet been given those names) and said that the consortium would buy 60,000 chip sets from Intel during the 36-month life of the contract, and possibly more thereafter. Monthly deliveries over most of the period would be 6,500 units.

The reliability goal was a failure rate of 0.01 percent per 1,000 hours of operation. Defects would be handled by one-to-one replacements.

The purchase price (which would be over and above the $60,000 development fee) would be $19.50 for the 4004 for the first 50,000 units, and $18.00 thereafter. The cost of the

4001 and the 4002 would be $13.00 for the first 50,000 units, and $12.00 thereafter. The comparatively simple 4003 would cost $2.15.

For the purposes of this narrative, the main point about the contract is that, inexplicably, it did not mention any delivery deadlines. The 36-month agreement appeared to extend from the time of the first deliveries, but just when those deliveries were supposed to begin was not stated. Nor was any penalty mentioned if deliveries were late.

This lack of specificity, and of non-performance consequences, would come to haunt the project, as it ended up sitting dormant for months, for no good reason. If the agreement with CTC was equally vague, that would help explain why that project, too, tended to languish.

CTC, in any event, successfully pressed ahead with the Datapoint 2200 without the chip. In the process it learned a lesson that was catastrophically wrong—that it was better off relying on its own processor design. For a decade that approach worked, but after that the firm found itself fatally isolated from the rest of the market.

Chapter 6

The 1201 Chip: Spring 1970 Through 1971

After CTC made the chip development deal with Intel, it returned to the task of trying to remain afloat while selling the Datapoint 3300 terminal, while at the same time developing, manufacturing, and seeking market acceptance for the new Datapoint 2200. As will be shown in later chapters, market acceptance of the 2200 would take care of itself—but everything else was a struggle.

At Intel, the chip project literally took on a life of its own—a charmed life. Its subsequent treatment by CTC and Intel (and also, as will be shown, by Texas Instruments) should have guaranteed that it never saw the light of day. But fate seemed to have other plans. Apparently, the idea behind it was so compelling that it was able to overcome all the snags that it encountered.

The first snag was taking form even as the development agreement was being drawn up between CTC and Intel. The cost of TTL components fell dramatically in 1970—by factors of more than ten in some cases. This left no compelling financial rationale for a CPU chip, although the need to save space and reduce heat remained important.

Meanwhile, the recession continued raging, so that developing a chip that did not promise an immediate return on investment (especially in the face of the falling prices of board-level TTL alternatives) must have seemed like a wild luxury.

Even before economics became a problem, the project was off to a bumpy start. At Intel, the project to develop the CTC chip shared resources with the Busicom calculator processor chip, later named the 4004. Its progress did not bode well for the CTC project.

Recapping Chapter 5, a Japanese consortium called Busicom had approached Intel in June 1969 to make the chips for a scientific calculator, and Intel had proposed replacing the original design, which involved about a dozen chips, with a four-chip approach involving a microprocessor. The processor had a data path only four bits wide and a limited memory capacity, making it suitable for a calculator (since numbers are usually encoded with four bits, and since only limited memory was needed) but unsuitable for a general-purpose computer. The delegation had involved engineer Masatoshi Shima, who returned in September to consult on certain details, and planned to return in April to check the final design.

In the meantime, after meeting with Vic Poor near the end of 1969, Stan Mazor at Intel went to his boss, Marcian "Ted" Hoff, and wrote an outline of what they thought a Datapoint 2200 chip should be like. After Poor sent them the manual for the Datapoint 2200 instruction set, Mazor and Hoff did a formal proposal.

"CTC's insight was that since they were getting shift registers on the cheap and since we (Intel) like to make them and they are already using them, why not reinvent the computer with rotating memory? Let's use the shift register as main memory," Mazor explained four decades later. "That was their insight, their invention. It was before there were RAM chips, only core RAM. Normally you had a 12-bit or a 16-bit computer, but they decided to build a one-bit computer, operating one bit at a time, to greatly reduce the circuitry and the cost. To the programmer it looked like an eight-bit machine but it worked on an eight-bit number one bit at a time." But the computer would have needed a way to return from subroutines—Mazor likened the process to hanging up after a phone call and returning to what you were doing before the call. A 64-bit RAM chip that Intel was making could do the job if it were paired with an address router, which led to the original conversation with Poor, Mazor said.

But when Mazor wrote the proposal for the CTC chip, he decided not to follow CTC's path, and make the chip an eight-bit device instead of a one-bit device.

"We figured it would be easier—so we thought—to build a parallel instead of a serial processor. I don't know if it really was since we never did build a serial processor. But with a serial processor you had to track what bit you were on, reassemble the bits into a byte at the end of the process, and keep everything synchronized. We concluded it would be easier to go parallel."

As noted, the details of the business deal between CTC and Intel to develop a true computer chip (history's first) are lost. But the basic business arrangements must have been settled by March 9, 1970, since on that day chip designer Hal Feeney began working for Intel, and was handed the CTC project.

Work Begins

Feeney had worked for the previous three years designing chips, mostly for military applications, in the California design office of General Instruments, but was laid off when that company was hit by the 1970 recession. He heard about the Intel opening through a friend and was almost immediately hired there.

The custom CTC chip was called the 1201.[56] Feeney recalled that the first 1 meant it the chip was to be made with the p-channel MOS process, the 2 meant it was a random logic chip instead of a memory chip, and the 01 meant it was the first product in that series. It is also described as Intel's first random-logic chip using silicon-gate technology.

Let's pause to explain what those words mean. P-channel MOS means a metal-oxide semiconductor using p-channel doping, or additives to create positive electrical polarity. At this writing, modern semiconductors use both positive and negative doping, and are called complementary metal-oxide semiconductor (CMOS) devices.

Meanwhile, "logic" in electronics does not refer a formal approach to reasoning, but a system for specifying what response a circuit should make to specified inputs so that the circuit can, in the end, perform a specified task. "Random logic" means that circuit it not based on the predictable kind of logic used in memory chips, whose working circuits are designed once and then replicated across the chip. "Silicon-gate" technology was then the latest generation of chip fabrication technology, replacing the previous metal-gate process that used aluminum electrodes. Silicon-gate technology produced circuits that were three to five

---

56 The 1201 was later given the name 8008, as will be subsequently discussed.

times faster and used half the chip space as the previous metal-gate technology.[57]

Feeney was immediately handed a nine or ten-page document, written in pencil by Mazor, Hoff, and their team, specifying the functions of the chip. "The most important part was the listing of the instructions, most of which were very similar to those of the Datapoint 2200," he recalled. "Initially, I was supposed to implement the logic, using such things as AND and OR gates, plus flip-flops, memory cells, and instruction decoders. All these would have standard implementations into actual circuits."

Again, let's explain what these words mean. The computations performed by digital computers depend on circuits that perform binary (also called Boolean) logic. Such logic involves the manipulation of ones and zeroes, which the circuitry expresses as on/off or high/low voltages. There are several basic binary logic operations, such as AND, NAND (i.e., Not-AND), OR and NOR (i.e., Not-OR). The circuitry that performs a specific logic operation is called a gate, and usually has two input lines and one output line. Typically, the binary operation will define what the output line of the gate will do in response to the state of the input lines. For instance, the output line of an AND gate will be a zero unless both input lines are ones, in which case the output is one. With a NAND gate, the output will be one unless both inputs are ones, in which case the output is zero. Each variety of gate has a standard implementation in terms of the layout of the transistors, connectors, and other circuit elements needed to perform the function of that gate. Therefore, the layout of the circuitry is determined by the logic gates. Each gate takes two to a half-dozen transistors. The chip had about 3,500 transistors.

(A transistor, to take it to another level of detail, is the basic building block of computer circuits. A transistor lets current on one line control whether current can flow on another line. For instance, if a transistor were a capital T, current could only flow across the top cross-arm if there was current on the upright arm. Transistors are often referred to as amplifiers, since the current across the cross-arm can be proportional to the current on the upright, and that's how transistor radios work. But in computer circuits the current is all or nothing.)

Feeney later left for a while for his April 1 marriage, and when he returned he found that Italian immigrant Federico Faggin[58] has been hired to work on the 4004 chip. Actually, it would have been named the 1202 chip (forever giving the CTC chip precedence) but Faggin convinced Intel's management that processor chips should be given distinctive names. The 4004 name was a play on the chip's four-bit data path.

Faggin, then about 27, had a doctorate from Italy and had come to California in February 1968 to work for six months as an exchange engineer for Fairchild Semiconductor. He was enchanted by California's springtime, in comparison to foggy, cold Milan in that season. After Fairchild sold its interest in the Italian firm that Faggin had been traded from, he became a full-time employee on July 1, 1968—the same day that Bob Noyce and Gordon Moore left Fairchild to form Intel. Working on the development of chip-making processes, Faggin felt

---

57 Background on silicon-gate technology is from the Web site of the Computer History Museum, http://www.computerhistory.org/semiconductor/timeline/1968-SGT.html, assessed June 30, 2009.
58 Faggin is pronounced Fah-JEEN. Information about him is from an interview with the author in 2009, with background from a Computer History Museum interview conducted in 2004, and a group interview concerning the history of the 8008 conducted in 2006.

frustrated—he saw himself as a chip designer. So eventually he joined the stream of Fairchild employees moving to Intel.

Faggin was not told the state of the Busicom project when he was hired, and only understood that it was "challenging." On his first day, April 3, 1970, he learned that the Busicom engineer, Matsotoshi Shima, would be flying in from Japan the next day to check on the progress of the design. Faggin then looked into the matter, and found that there had been no progress. There was a formal description of the block architecture, without much detail, and some trial circuit diagrams that later proved to be useless. There were no logic diagrams.

"It was a staffing issue," Mazor said later, about the delay. "There were two engineers in the department, Hoff and myself. We had the architecture but they had not hired an engineer to build it, and that is why the 4004 was late."

Indeed, it appears that no work had been done on it for five months. There was no reason to expect Shima to remain calm when receiving the news upon his arrival, and he didn't. So the meeting with Shima the next day—Faggin's second day on the job—was somewhat stressful, but Faggin eventually calmed him down. After about a week a reconciled Shima got Busicom's permission to stay in town and help with the logic design, Busicom having decided it would be too much trouble to find another vendor.

Faggin began working feverishly, doing the layout in parallel with the logic design, instead of finishing the logic design and then doing the complete layout. He found that his bosses assumed that he could lay out the CPU chip—the 4004—in one month, based on the time it usually took to do a RAM chip. But the trick to doing a RAM chip was to find a way to arrange the necessary number of identical memory cells onto the die. With a CPU, the designer worked with logic gates rather than memory cells, and each logic gate would likely be different from its neighbor. The job was vastly more complex, in other words. Faggin was expected to have working samples of all four chips by September, which was clearly not possible. (Faggin remembered feeling jealous of Feeney, since the latter had only one chip to design, and therefore would likely produce history's first processor chip to reach the market.)

As for Feeney, he continued pressing ahead on the 1201 project for CTC, designing it to fit into a package with 16 pins, that being standard at the time for Intel chips. He finished the first full functional specification for the chip on June 25, 1970, with the first revision completed on July 20.

The Hiatus

And about then, work stopped on the 1201. Feeney managed to do the block diagram for the chip in October, but there were no additional milestones for three months, until January 25, 1971, when he did revision 2 of the functional specification, adding some instructions.

Of course, by July, 1970, CTC had completed functional prototypes of the Datapoint 2200, had shown them at a trade show, and was ramping up for production, which was to begin in November. Thanks to its use of a CPU that occupied an entire circuit board, CTC could get by without the chip, although it had plans for using the chip, if it came along.

Meanwhile, a recession was underway, and CTC's lease-based revenue had not yet exceeded its expenses, meaning it was losing money on a daily basis. Mazor, for instance, remembers that the 1201 was put aside after CTC stopped paying its bills for its shift register memories from Intel.

"Other people recall it different ways, but from my perspective it was not called off," Feeney said. "I was told that there were higher priority things that we needed to do, and to put it (the 1201 chip) on the back burner." He thereafter spent a lot of time on the 4004 and its companion chips, but a lot of that work was directly applicable to the 1201 project, such as the development of testing and debugging tools and techniques.

"It really felt like the same project" he recalled. "We were learning all the way through the process." The main lesson, he recalled, was the use of the "buried contact" chip design technique that Faggin had pioneered at Fairchild, that let them add connection points between layers of circuitry. This was first used in the 4004, and without it the 1201 would not have fit within the available chip package, Feeney said.

However, Faggin remembered that the 1201 project was at a standstill even before CTC pulled back. "Feeney did not have the knowledge to really develop a chip—there was no methodology at Intel to make random logic circuits. Texas Instruments had made random logic before and had the tricks and tools, but Intel was a memory company and did not have anything. Memory involves repetitive designs, where the cells are repeated over and over again. You beat the hell out of the cell design to make it optimal under all conditions, and you can spend your time since you have only one thing to worry about. With random logic, each gate is different, and you have to design them very quickly.

"And from the start, the (new) silicon-gate technology required a different way of laying out the circuits, so the project did not go anywhere," Faggin said. "Feeney was reassigned and eventually ended up working for me. In January 1971, after the project was restarted, Feeney was working for me and did the project using the methods and tricks that I developed for the 4004."

But what really annoyed Faggin was that Intel insisted that the 1201 be housed in the same 16-pin package that Intel was using for memory chips. The use of 16 pins was sold as a virtue since a new package did not have to be developed. But with only 16 lines connecting the chip to the outside world, the chip did not have enough connectors for eight separate control lines representing the data path, and the 13 lines needed to address 8K of memory. It ended up using eight lines that ran in both directions, and the system needed extra chips— as many as three dozen—to sort out the signals that shared those lines. Since the signals were time-multiplexed, meaning the various types of signals had to wait their turn to use the pins, operations were slowed down even further.

"The number of pins was a dumb idea but I was stuck with that decision," Faggin complained later. "Intel was ignorant of what it meant to have a microprocessor. I wanted to have 40 pins, to have more parallel activity and an interface to memory. There was a dogma around 16 pins, but it meant that you had to put random logic around the chip, and that was dumb. So much for a single-chip microprocessor."

However, during the 1201 project hiatus that ensued after July 1970, Intel found that it needed an 18-pin package for its latest RAM chip. After work resumed, the 1201 was able to adopt that package and get two more pins. "They could not get the 1103 (RAM chip) on 16 pins since they had to put a reverse bias on the substrate to avoid problems with retention time," Faggin said. "I remember the day that it became clear that they had to put a reverse bias on the chip. It was like somebody had died. In fact, it was like all the top management had died in a car accident. There were long faces around Intel for days when it happened that we had to use two more pins."

In the end they were able to invoke 14 address lines, letting them use 16K of RAM. At the time, this was considered a generous amount.

(At this writing, CPU chips use about a thousand pins, covering the back of their square, flat packages so that they resemble small grooming brushes with stiff, regimented bristles. In 1970 the pins were arrayed along the perimeter of the long sides of the rectangular chips and were called dual in-line packages, or DIPs. The micro-circuitry of the chip occupied only a tiny square—the die—in the middle of the package, and the rest of the package just held wires that connected outputs on the die to connector pins along the edge of the package. The pins would fit into precisely placed, metal-plated holes in a circuit board, where they would be soldered. These holes would be connected to other components on the board through wires etched into the board.)

Between October and Christmas, Faggin managed to get one after another of the four Busicom chips designed and laid out, and then got sample prototypes made. The 4004 itself was the last one through the design pipeline, with the first sample being produced around Christmas, 1970. It didn't work—one of the circuit layers had been omitted.

This was corrected and a second attempt in mid-January, 1971, succeeded. The design work was finished on the 4004.

"As soon as we had functional samples of the 4004, all my time went back to the 1201. At that point management wanted the 1201 to come out as quickly as possible," Feeney recalled.

The reason was that, in early January, Intel was approached by a Japanese firm, Seiko, who wanted the 1201 chip for their own scientific calculator. Feeney remembered that they wanted to make something equivalent to the Hewlett-Packard 9100, a 40 pound desktop electronic calculator introduced in 1968 that could hold 196 programming steps. It had a three-line screen and cost upwards of $5,000. (Science fiction writer Arthur C. Clark was given one by HP and later said that he was thinking of it when he gave the name HAL 9000 to the mutinous computer in "2001: A Space Odyssey.")[59]

Design work proceeded through various steps, with the layout being finished on July 23, 1971. The 1201 was the last Intel chip whose layout was drawn by hand. Half the chip could be done at a time, at a scale of 500 to 1, on a table the size of a ping-pong table.

---

59 The reference is from http://hp9100.com/, accessed July 10, 2009.

Tom Thumb?

Even if the 1201 chip was not out yet, and even if it was falling farther and farther behind, it was exciting imaginations—or, sometimes, fantasies.

In 1971 Herb Baskin[60] was a professor of computer science and electrical engineering at the University of California at Berkeley. Having assembled a team of engineers to start a company involved in high-technology for hotel security, he was suddenly approached by a promoter who needed designers to flesh out an idea he had.

He told Baskin he had plans to build a very small computer, code-named Tom Thumb, with a keyboard, a display, and a one-chip processor. He said that he had acquired a license to use an Intel one-chip processor, but it was not available yet. However, the technology was already in use by a company in San Antonio that Baskin had never heard of, called CTC.

To learn about the technology, Baskin visited CTC with the promoter, and met with Vic Poor. Baskin recalled that Poor was leery of the promoter.

"Poor almost invisibly slipped me his business card and told me to call him sometime," Baskin recalled.

Nothing came of the Tom Thumb project. The promoter's "company" consisted of little more than his girlfriend posing as his secretary. (How he could have had a "license" for the 1201 chip is another question, as it was to be sold on the open market.) Soon, Baskin decided to accept Poor's offer, and called him.

Poor hired him as a consultant, initially to examine the suitability of the Datapoint instruction set (presumably the expanded version for the upcoming Datapoint 5500 processor.) Baskin, meanwhile, remained a professor at UC Berkeley.

"I was computer architect by self designation, although I had graduate students who were smarter than me and had better insights into instruction sets," Baskin recalled. "We combed through the instruction set proposal, and advised Datapoint (CTC) concerning our evaluation of the instructions. I don't think any changes came about. Instead we gave them confidence that it was suitable enough. But their main source of confidence was their ability, through brilliant programmers, to create impressive software components."

After that, Poor retained Baskin as a consultant to pursue a number of projects, mostly involving software development. He was invited to attend monthly planning meetings in San Antonio, and Poor asked him to, basically, show the flag for Datapoint in Silicon Valley.

"They wanted me to remain a professor and serve as their link to the West Coast and the academic community," Baskin explained. "They had almost no one in San Antonio with credentials in the computer science community at the faculty level. I was their small foray into that area."

One of his first projects was a floating-point math package, demonstrating the kind of math used in scientific calculations. Shown at a trade show, it caused suspicious attendees to repeatedly look for some connection between the Datapoint 2200 and a mainframe.

---

60 Baskin was interviewed in 2009.

Poor also asked Baskin to look after the progress of the 1201 chip. This involved occasional visits to Intel, and once Baskin found himself chatting with Intel's president, Bob Noyce, in the evening after the day's work was wrapped up.

Baskin noted that the 1201 chip was way behind schedule, and Noyce actually explained why.

"He confided in me, as a fellow West Coaster, that he did not believe that it was the right way to go, to accumulate more and more logic on a chip. The future was in putting more and more memory on a chip," Baskin recalled.

Noyce made it clear that the large shift register contract with CTC (the one Bob McDowell had stumbled into) had been a godsend. It had come at a time when Intel was out of money and unable to raise more. He saw San Antonio as a technological boondocks, capable of producing only a quiet Teletype (the Datapoint 3300.) If Intel was dragging its feet about the CPU chip development project, it was because that project was not where the future was.

"They used the money (from CTC) for memory chip development," Baskin concluded. "Noyce finished the chip project only out of a sense of responsibility."

The Texas Instruments Interlude

CTC's relationship with Intel suffered another setback with the arrival of the June 7, 1971, issue of Electronics Magazine. Stan Mazor at Intel vividly recalled opening it and, on page 36, coming across a story headlined "New chip turns terminal into standalone machine." It contained news from Texas Instruments that TI evidently intended as a bombshell—and for Intel it certainly was.

The nine-paragraph article, with a picture of a chip, said that Texas Instruments would be enhancing the functioning of the Datapoint 2200 with a new processor chip, "a monolithic MOS circuit containing about 3100 transistors."

It described the 2200 as the "first integrated remote terminal to provide local computational capability, reducing the load on the master computer in supervision and routine data processing." The article described the 2200 as an eight-bit computer whose memory capacity of 8,192 bytes gave it the capability of a DEC PDP-8. But it said that TI was developing an LSI (large-scale integration) chip for CTC for use in a new version of the 2200 that would come out in August of September, turning the 2200 into "a complete computer that doesn't have to be connected to a time-sharing system." Like the 4004, it would use RAM memory chips that were specially developed for that processor, and which would not necessarily be compatible with anything else. A TI executive was quoted, by name, as saying that chip, with 24 pins and a die size of 212x224 mils, was not uneconomically large—at least not for TI.

The article then cited an unnamed CTC spokesperson as saying that the component cost of the new chip and 2K of memory, "enough for many source data applications," would only be about $100. And, that the new 2200 would be able to support 64K of RAM instead of 8K, but the memory space above 16K would be devoted to ROM which would contain Basic[61]

---

61 Basic (or BASIC) was and is a fairly simple computer language used to teach programming in schools, but has also been used to produce commercial software.

or some other programming language. The new 2200 would probably lease for $200 per month, which was a steal compared to time-sharing port charges of $600 to $1,000 per month.

In hindsight, there are some odd features about this article. For instance, at the time CTC was, indeed, readying a new version of the 2200, but it was not exactly expected to be unveiled in the next two or three months. In fact, CTC was then scrambling to avoid bankruptcy, and the Datapoint 2200 II would not appear for a year. As for TI's chip allowing the 2200 to become a complete computer that did not need to be connected to a time-sharing system, hopefully someone at CTC could have told that writer that CTC's customers were already using the 2200 that way. Someone at CTC might also have told the writer that comparing the one-bit 2200 to the 12-bit PDP-8 was something of a stretch.

Especially startling to modern eyes is the announced plan to put Basic in ROM in upper memory. Doing so freed expensive RAM from having to hold the programming language and the operating system, and it became a common feature in early personal computers—about six years later. But CTC (or TI) did not originate the idea, as the Apollo program in the 1960s had used guidance computers with software stored in ROM. Basic, meanwhile, had originated at Dartmouth University in 1964, and was well-known by 1971.[62]

Of course, to anyone on the Intel design team for the 1201, the article would have been very disturbing—TI appeared to be bragging that it had already done what Intel had been trying to do for more than a year, make the world's first microprocessor for the Datapoint 2200. Unbeknownst to Intel, CTC had also been working with Texas Instruments.

"It was not hard to see that it was the 8008 done way ahead of us," Mazor said. "We were just blown away—it was like finding that your partner is cheating on you. We were amazed and unhappy.

"Normally, when I give a customer a chip specification I have them sign a non-disclosure agreement, saying this is something we are working on and you are not to release it to a third party," Mazor said. "I never did that with CTC, since the specification was custom with them. And I was young and naive and it never occurred to me to do that. But, unknown to me and Intel, my handwritten proposal went to TI, and they started a crash program. When we put down the program due to CTC's financial problems, TI did the opposite, and used automated tools to get the design done faster than we would have otherwise."

He was able to console himself after examining the chip's specifications, however. "By our reckoning the TI chip was about twice as big as the chip we were planning to make. Although they beat us, and we were amazed, surprised, and disappointed, we thought their chip was ridiculous. It was much too large and therefore too expensive, since a chip twice as large costs four times as much. It was like something made with Craftsman tools versus something made with Lego blocks."

As for how this happened, Poor explained that in early 1970 CTC was buying parts from both Intel and TI, and the TI salesman wanted all that business for himself. The salesman heard about CTC's deal with Intel concerning a processor, and made a presentation, touting TI's

---

62 Michael Fischer, personal communication, 2011.

processor expertise, and proposing a three-chip processor from TI. Its instruction set was incompatible with the 2200, and Poor turned it down, pointing out that Intel had agreed to implement the 2200 instruction set. So the TI salesman took the bait and agreed to do the same thing.

"They had just set up a MOS facility in Houston and it was the kind of thing that they wanted to do," Poor recalled. "So we gave them the same specification we gave to Intel, but we did not commit ourselves to another NRE (non-recurring expense, or development fee.) We told them what it would take to be a second source, although we did not yet have a first source."

In other words, he told TI that they were free to try, but they were on their own—CTC would not fund the effort. Poor recalled that the deal must have been made in March or April of 1970, about the time that work on the 1201 chip was starting at Intel.

Jonathan Schmidt's archives contain a copy of a typed TI document titled, "System Definition of a One-Chip MOS/LSI Microprocessor." It was dated June 23, 1970.

A memo in the Roche family archives is presumably an early draft of the press release that triggered the Electronics Magazine article. It may have been sent to Roche for his approval, although it contains no marginal notations. It was written in pencil on engineer's graph paper, dated February 24, 1971, bore the imprint of a Texas Instruments rubber stamp, and was titled "Description of CTC DP2200 System."

It stated that TI had developed a two-part solid-state component kit for CTC's Datapoint 2200, consisting of processor and memory devices. The processor was a "complete miniature computer" and represented a "breakthrough in integrated circuit design" with 3,078 transistors on a bar of silicon 0.20 inch square. Prototypes were being evaluated with production scheduled to begin in April.

But it was all for nothing. Sometime in the summer of 1971 the TI chip—the first of billions of x86 processors that have since been made—was sent to CTC for evaluation. And what happened shows that the story in Electronics Magazine was largely wishful thinking, since the CTC engineers couldn't get the thing to work, and rejected it. The situation may have run its course by the time the article appeared.

Future Datapoint executive David Monroe was the technician assigned to test the TI chip. He recalled that it was delivered as a two-chip set—the processor chip, and a custom RAM chip designed to work with the processor. As far as he could determine, the processor chip did work, as it responded correctly to what input he was able to give it. But it could only run programs when paired with its custom RAM chip, and he could not get the RAM chip to work.[63]

"It had such a narrow voltage and temperature operating range that it was not usable—it was too fragile," Schmidt recalled. "There was a recession at the time, TI shut down their MOS factory in Houston, and that project was killed."

So there it ended—except that, on August 31, 1971, TI engineer Gary Boone filed a patent

---

63 Early RAM chips were often flakey. Monroe recalled one instance where a RAM chip worked fine in the lab, but periodically failed in the field. It turned out that in the lab it was exposed to light, and that made the difference.

application for the chip. Moreover, when specifications of the patent were published, Mazor became convinced that TI had used an early version of the functional specification for the chip that he wrote. Somehow, it has been passed to TI.

The suspicions of betrayal, coupled with TI's attempts to enforce the patent in the early 90s, generated a soap opera whose echoes continue at this writing, coloring the responses and attitudes of many of this story's participants.

Meanwhile, TI was not the only party to file a patent on the microcomputer chip. The patent landscape will be examined in more detail in Appendix A.

First Fruits

Meanwhile, Intel had begun making the 4004 chip sets and delivering them to Busicom. Almost immediately, in the spring of 1971, Busicom began running into financial problems due to brutal competition in the calculator market. (It went bankrupt in 1974.) Intel agreed to lower the price of the 4004 chip set in exchange for the right to sell the chips on the open market, to non-calculator firms. The deal did not represent any pre-existing desire on the part of Intel to sell processor chips—apparently, exclusivity was the only bargaining chip that Busicom had, and Intel accepted it. And, of course, there was the realization that if Busicom went under, Intel's one customer for the 4004 would go away. Obviously, it would be better for Intel if the product was non-exclusive.

But this one small contract renegotiation set the stage for everything that followed. Intel would become a merchant supplier of processor chips. When Intel succeeded in that business, other chipmakers followed. The end result was that microprocessor chips would not be captive products of established computer vendors, but would be available to anyone interested in trying to create a computer—or a computer company. The computer industry would be transformed from a narrow, arcane market controlled by a handful of large, specialized firms, into a thriving ecosystem that no one could claim to control, but which touched every aspect of the economy.

But that lay in the future. In the summer of 1971, Intel remained reluctant to market the 4004, since it was so different from its other products. But then Intel hired a new vice president of marketing from Texas Instruments. He saw things differently, and Intel began advertising the 4004 in the November 15, 1971, issue of Electronic News.

"Announcing a new era of integrated electronics. A micro-programmable computer on a chip!" trumpeted the ad. It included a drawing of office workers with a terminal and a printer, in front of a bank of mainframe cabinets. On closer inspection, the "cabinets" turn out to be refrigerator-sized 4004 chips standing upright. Of course, they have 16 pins.[64]

At one point at a computer conference soon after the release of the 4004, Mazor found himself confronted by an indignant engineer demanding to know how Intel could dare to say that it had a computer on a chip. Mazor handed him a data sheet and let him read it. "God it really is a computer!" gasped the doubter.

---

[64] The ad is shown at http://www.intel.com/museum/archives/4004.htm, accessed July 22, 2009.

Initially, buyers had to pledge not to produce anything that would compete with Busicom's calculator, but that limitation was later lifted.

The 4004 had 2,300 transistors on a die about the size of a fingernail, or one-eighth by one-sixteenth inch. The circuit geometry was 10 microns, meaning that the etched lines that served as wires were ten microns wide. A micron (designated μm) is a millionth of a meter, or one-thousandths of a millimeter. By comparison, the cross-section of the average hair is about 100 microns. At this writing in 2009, the leading-edge circuit geometry of microprocessor chips was 32 nanometers. A nanometer (designated nm) is one billionth of a meter, or one thousandth of a micron. So the circuit size had been reduced by a factor of more than 300. The clock speed of the 4004 was about 740,000 cycles per second. At this writing, processor clock speeds topped out at about 4 billion cycles per second, or more than 5,000 times faster.

About the time the 4004 came out, Intel was also able to show working samples of the 1201 chip to CTC. Exactly when is difficult to pin down but sources tend to agree that Intel was a year late with the chip. Since the original deadline was probably in late 1970, being a year late would have put the event in late 1971.

What happened next would determine the future of Intel and CTC, plus the computer industry as a whole—and, arguably, the entire industrialized world.

But to understand what happened, we need to lay out what CTC was doing during this period.

Chapter 7

CTC Makes the 2200: 1970-1972

Vic Poor, as noted, began working for CTC in December 1969 to design the Datapoint 2200. He met with Intel around Christmas to discuss turning its processor into a chip, initially called the 1201, but design work on the 2200 proceeded without that chip.

CTC had begun shipping CTC's initial product, the Datapoint 3300 terminal, three months earlier. Poor recalled that Phil Ray liked to display a report from a market research firm indicating that the 3300 had no future and that no one should invest in it. The market was not big enough, the 3300 did not have enough features, and it sure didn't have any future. The report's estimate of the total market was less than the number that CTC sold in the first 60 days. This meshed with what Poor knew of market research. At Frederick, his approach had been to build a prototype and see if anyone got excited about it.

Jonathan Schmidt followed Vic Poor from Frederick Electronics in Maryland to CTC in San Antonio. He recalled visiting in late 1969, and starting full-time in January 1970.

"I thought Texas was desert and cowboys," he said. "But San Antonio had a classical music station and a symphony. It was dry, meaning you could not get beer or wine except in a private club—which would sell you membership for a dollar when you walked in. You could not buy anything on Sundays except food and gasoline, unless you declared it was an emergency purchase.

"The wine was flowing when I got there, with fancy dinners at clubs, with big parties by CTC, or rich people who had become friends of Gus and Phil. The people there worked and played hard. It was a different environment from Frederick (Maryland), which was a little country town with a Dairy Queen," Schmidt said.

CTC had already moved to its new custom-built facility in northwest San Antonio, near what is now the corner of Wurzbach and Datapoint. The building would be expanded several times while the company likewise sprawled into about 40 other buildings scattered through the city.

When Schmidt arrived, the processor was a breadboard the size of a conference table, and was not functional yet. Schmidt was called on to write a simulator for the 2200 on a Hewlett-Packard computer, so that they could start writing software for the 2200 before the first one was running. (Pyle wrote an assembler before returning to school.)

There was no time to wait because, Schmidt recalled, Roche wanted to debut the Datapoint 2200 at the American Bankers Association convention in San Francisco in April—only four months away. The result was 100 days of pressure.

"It was a ruse to get the engineers working 20-hour days," Schmidt recalled. "Everything was very wild. People would work to 10pm, go to bar until 2 am, and show up at work at 9 am. Sometimes we would work all night."

Roche wanted demo programs that emphasized intelligent data entry. The approach followed that used by IBM mainframe terminals, where the screen is divided into fields with fixed attributes, including input fields whose contents would be transmitted to the mainframe. With the demo software, input fields that were supposed to contain names would not accept numbers, and fields for numbers would not accept alphabetical characters. Once entered, data could be loaded to a tape and then displayed again. (In real-world use, their contents would then be transmitted to a mainframe, in one short session.) Schmidt recalled finishing them in his hotel room just before the start of the convention.

There were also simple games written in time for the April convention. (These got more attention than the data input demos, Poor recalled.) The only one anyone clearly remembered was a slot machine simulator. There may also have been a Teletype emulator, but the small staff would not have had time to do much else.

All the programs self-loaded from the tape drive, one program per cassette, there being no operating system yet.

The Machine

Weighing 47 pounds, the Datapoint 2200 was intended to occupy the footprint of an IBM Selectric typewriter, meaning it was 18.5 inches wide and 19 and 5/8 inches deep. Thanks to the half-height screen, it was only 9 and 5/8 inches high.

As discussed, the half-height green monochrome screen was intended to mimic the aspect of an IBM punch card, and was 7 inches wide and 2.5 inches high, capable of displaying 12 lines of 80 characters. (Its half-height cathode ray tube had to be specially produced for CTC, increasing development costs.) It would display all 96 alphanumeric characters of the ASCII code as were then standard, meaning it could display both uppercase and lowercase characters. (Many terminals at the time followed the example of the Teletype and could only display uppercase letters.) There was no provision for graphics.

When punch card emulation was forgotten, the lower height was recast as a feature in its own right—the shorter 2200 was less intimidating to secretaries than the towering 3300, which was 14 inches high.

The built-in keyboard had rollover similar to the 3300. It had a standard QWERTY keyboard with a numeric keypad to the right, as commonly seen on today's PC keyboards. For the sake of feedback, the system could optionally produce a click when a key was pressed. On the right end of the keyboard was a column of five large system keys: RUN, STOP, KEYBOARD, DISPLAY, and RESTART.

RESTART caused the rear tape drive to rewind and then load the first item from the tape into the main memory, and begin executing it, be it the operating system or a stand-alone program. STOP halted processor operations, and RUN resumed processor operation. The actions of KEYBOARD and DISPLAY were controlled by the software.

The processor could address 8,192 bytes of internal memory embodied in shift registers. The processor itself was embodied in about 120 medium-scale integration (MSI) chips, consuming at least a full board. The designers had a backup plan for using the Intel 1201 chip, but, as discussed, did not wait for it.

There were two cassette tape drives in the top of the unit, both under full software control. The rear one was intended for the program storage, and the front one for data storage, at least at start-up. Each tape could hold about 130,000 bytes. Data could be loaded at a rate of 350 bytes per second, which was 35 times the speed of the paper tape it replaced. Rewinding a tape from the end took about 40 seconds.

The 2200 was rated for 180 watts. It used an early example of a switching power supply[65] instead of a transformer, which would have been too large for the cabinet. Reportedly, it often proved inadequate. The tape drives proved to be balky, and their drive belts were prone to wear. Static electricity would knock a 2200 off-line. One early customer told Poor, "I'd give up on the damn thing if it just didn't look like it ought to work so well."[66] In other words, its sleek design won it pity.

During the summer, preparation began for mass production—and the hiatus began on the development of the 1201 chip at Intel.

The 2200 prototypes were also shown at the Fall Joint Computer Conference in Houston, and Poor remembered that people with IBM name tags were constantly in the booth, showing rapt interest in the machine.

"Scare Hell Out of Your Secretary"

Frassanito recalled that the marketing plan was to offer a selection of self-loading programs, each with a different application. There would be a basic line-by-line text editor (since nothing fancier could be done in 8K), data entry applications, the 029 emulator, and other terminal emulators.

Immediately after the machine was shown successfully in San Francisco in April, CTC's managers began planning a national advertising campaign to accompany the release of the product, expected later that year. Unable to find any local advertising talent he considered even second-rate, Frassanito went to New York, where the Swinging 60s had turned advertising into a glamour profession. He had hoped to sign up the agency that produced the Volkswagen Beetle ads that were turning heads at the time, but got turned down. Things also seemed to go equally badly when he initially approached George Lois of the firm Lois Holland Callaway. Frassanito remarked about the surname Lois being Italian, and was informed (with much use of the vernacular) that the name was Greek.

But Lois (now famous for, among other things, naming Lean Cuisine, for creating the "I Want My MTV" campaign, and for doing 40 Esquire covers) agreed to come to San Antonio where he hit it off with Roche and Ray. He proposed that what CTC needed was something outrageous, to ensure that its unprecedented product got the attention it needed, Frassanito recalled. Instead of relying on dull text, Lois sought to grab the attention of the reader, and then convey the central idea with as few words as possible.

The result was a full-page ad with the headline, "Scare hell out of your secretary. Get her

---

[65] In electronics, a "power supply" is not a source of electricity, but the component that converts the AC power from the wall receptacle to the lower-voltage DC power that the digital circuits use. A switching power supply does this by flipping the power on and off at a high rate, so that the power that gets through is what the circuits require.
[66] Datapoint vanity history of 1982.

a computer." The ad was dominated by four rows of five pictures each showing a well-groomed young woman (presumably a secretary) reacting to the sudden appearance of a Datapoint 2200 on her desk. Nineteen of the pictures show her in various stages of shock and fright, with her hair as a fright wig.

In the twentieth, she's happy and confident. "About 45 minutes after the handsome thing is on her desk, she'll be an expert," promised the subhead.

A copy block under the second row of fright-wig pictures continued the theme. "Look at the Datapoint 2200. It ushers in the sensible age of computers. She'll enter data directly from source documents. She'll verify it on the CRT screen. And it's transmitted with no other human involvement. (No mistakes either.) She doesn't need any other equipment. It's a friendly terminal that talks to her. Guides her. Walks her through the job. And without a peep or noise. It's the first computer that doesn't add to office turmoil."

An adjacent copy block was titled, "Some slightly technical information about the 2200," and below it were seven numbered items.

"1. It has a programmable memory of up to 8192-8 bit words." (Obviously, they meant "8192 8-bit words.")

"2. It takes a library of basic systems created by CTC." (By "takes" they must have meant "runs.")

"3. It enters data directly to tape, at the speed of light."

"4. It works with any data code. ASCII, EBCDIC, BCD, etc."

"5. Beautifully enough, no special training is required."

"6. Sleek and handsome. (Wait till you see it!)"

"7. Self-continued unit. That means you lay out nothing for supplemental units or auxiliary power."

The corporate logo was placed under the third row of pictures, with the motto, "Secretary's computer from Computer Terminal Corporation (the people who took the terror out of computers.)"

Taken as a whole, the advertisement obviously shows that CTC's management saw the Datapoint 2200 as a desktop computer, and hoped to sell it as such. The only application mentioned was intelligent data entry, but there was also point #2 concerning "a library of basic systems created by CTC." That referred to their previously mentioned plan to provide canned software packages for various desktop applications—but if they had used those words, no one at the time would have known what they meant.

Point #5, that no special training was needed, might seem like an exaggeration, but at the time a "computer" implied something like the IBM 360 mainframe, whose system documentation was said to consume six feet of shelf space. By that standard, the Datapoint 2200 really did require no special training. Meanwhile, mainframes often consumed thousands of watts and required special power adapters, which led to point #7.

After the ad was produced in the summer of 1970, CTC ran out of money. This is the same financial crisis that led CTC to stop paying its bills to Intel, resulting in the suspension of the 1201 chip project. Additionally, there was no money available to actually purchase space in magazines to run the ad, so it was put aside.

To Frassanito, the inability to run the ad was the biggest of the setbacks that CTC experienced at the time. National advertising would have greatly enhanced sales. And there would never have been any belief that CTC was selling programmable terminals—the ad called it a computer, and that was that.

(About a year later the ad was run in several magazines, but the marketing people that CTC had by then recruited thought it was successful mostly at being outrageous, if not embarrassing. They dropped it again, and also dropped Lois. A copy of the advertisement surfaced in the office archive when the author was working for Datapoint's product publicity department in 1982. The advertising manager shuddered when she saw it.)

First Sale

Working and living in the Minneapolis area, Dave Gust had worked for Control Data Corp. (CDC) in early 1960s as a junior engineer, having learned electronics in the Air Force. He later left to join an automated drafting startup with other CDC alumni. They were involved in a process that included coding schematics from a Teletype keyboard. They began looking for alternatives to the Teletype and found the Datapoint 3300.

"I got intrigued by the 3300," Gust recalled. When it became clear that the automated drafting firm was doomed he approached the local CTC sales office about joining them. They had an opening, so he hired on. It was 1970.

"We had four people in the Minneapolis office, counting the sales manager, but in the end they all left except me," Gust said. CTC literature at the time would list his home address as the Minneapolis sales office. "We sold a lot of 3300s, and it was the premier time-share terminal at that point. There were a couple of time-share companies in Minneapolis and we sold quite a few to them. Some of its features, like key rollover, represented a new concept. Its display was excellent—state of the art. It was pricey compared to a Teletype, but it was a clean, quiet terminal that I could put in clients' offices where they could do (phone) inquiries, like any other office equipment. We could also add tape and a terminal printer. It did not take too long at all before we were well known.

"But the business was really not enough to sustain everyone, plus they were beginning to clamp down on commissions. Also, there was turmoil because of uncertainty about the future, as the release of the Datapoint 2200 was lagging a little. The other people decided they could make better hay somewhere else.

"I took over the Minneapolis territory and started building it from there on. I ended up as the only person there in about 1971," he said.

After the Datapoint 2200 was released in late 1970, he began cold-calling. His computer background led him to believe that the machine had wider potential than the CTC marketing department seemed to accept. Actually, the machine struck him as being more powerful than some CDC computers he had worked with the early 1960s.

"The original idea was that we, CTC, would come up with some canned programs like terminal emulators," Gust said. "The buyers would use the tape cassettes for data collection. That they might go beyond the canned programs had not occurred to CTC at that time. The thought had not entered their minds to give the user the capability of sitting down and writing programs and doing some things on their own."

In his sales pitches, Gust always pointed out that the only cord coming out of the 2200 was the power cord—it did not have to be connected to anything else.

One of the companies that expressed interest was Pillsbury, a food conglomerate headquartered in Minneapolis. It had a problem whose answer might lie in the 2200.

Chicken Farms

The problem, Gust found, was that Pillsbury was running a number of chicken farms[67] throughout the South, and the payroll checks were mailed from Minneapolis. If there was a delay in the mail and the checks did not arrive by payday, the employees might not come in the next day. Obviously, it would be better to process the payrolls locally.

"No one else could offer them anything short of putting a mainframe in each chicken farm, which they did not want to do," Gust recalled. "There was the Data General Nova, and the DEC PDP-8, but those were rack-mounted boxes, and you had to have a control console and all that stuff. None of them were in a nice, attractive, user-friendly package. In terms of an actual, stand-alone device, there really was not any competition. I don't remember if attractiveness was an issue with Pillsbury, but I definitely remember that they were attracted to the fact that it was in a single, stand-alone box."

Gust recalled doing extensive concept selling with the Pillsbury buyer. "We were both on the same wavelength as far as the capabilities of the box. We could see that it had a processor and memory and a programming language, be it assembly language. There was the capability to sit down and write little programs to do a sort-merge (payroll) update," he said.

During the sales process, Roche and Ray flew to Minneapolis to talk to the Pillsbury people. "They were surprised at what Pillsbury wanted to do. They went back to San Antonio, sat down with Vic Poor, and said, 'We are missing the boat, as far as how we market it.' At that point they began looking at marketing it as a stand-alone device, and adding more peripherals and other things to make it a full system."

A preliminary copy of the 2200 programming manual was sent to Pillsbury on February 12, 1971.

Later, Gust flew to San Antonio and returned with one of the first 2200s off the assembly line, strapped in the first-class seat beside him. At Pillsbury, he demonstrated a small program that let him enter and record data.

---

67 A "chicken farm" is a place where chickens are raised in large numbers for slaughter. Some sources use the term interchangeably with "chicken ranch," but the latter was a Texan idiom for a bordello. Confounding the risk of confusion, sources agree that Datapoint did later have some "chicken ranches" as customers.

"I could show them that it was a real computer. They were won over and said that they wanted to do it," Gust said.

A Pillsbury computer manager got a training certificate from CTC dated March 22, 1971. The first end-user Datapoint 2200 was installed at a Pillsbury office in Minneapolis on April 7, 1971. It was used to program the payroll application, in assembly language. It used a printer directly attached to the 2200. Installations soon followed at four chicken farms, the first one being in El Dorado, Arkansas.[68] The payroll program would run over-night, and the checks would be printed out in the morning. The results would then be transmitted back to the headquarters mainframe in Minneapolis, to update the accounting files.

The Arkansas chicken farm represents the first known employment, by end-users, of what is now called a personal computer.

Pillsbury would eventually buy about 30 of the machines, and write a wide range of programs for them, such as for chicken feed formulation. Other, larger sales in the Midwest soon followed, affording Gust a comfortable living, he recalled, as no significant competition began appearing until 1972.

But while the Pillsbury account may have involved the first installation and use of a personal computer, it was apparently not the first sale to an end-user. Confusingly, both occurred in Minneapolis. Frassanito's archives include a souvenir copy of a sales agreement between CTC and General Mills, for 40 units, dated May 25, 1970—about a month after the first prototypes had been shown in public, and well before production began that fall. The first unit, an 8K model, was to be delivered as soon as possible with the balance in March 1971. "This is a conditional order based on a successful performance evaluation of the first installed unit," said the sales agreement. The evaluation period was to last 60 days.

While General Mills and Pillsbury were both in Minneapolis, they were then separate companies, the former acquiring the latter in the year 2000.[69] Presumably, CTC was not able to follow through in time to satisfy General Mills. Gust, who later sold hardware to General Mills, was unaware of any earlier sale to that conglomerate, and certainly of any installation that preceded the installation at Pillsbury. The document, therefore, is of interest mostly for the lease prices it shows: $168 per month for a 2200 with 8K of memory, $148 for a basic 2K unit, and $30 per month for a modem.

"Flummoxed"

But while the General Mills sale passed into obscurity, the Pillsbury sale lived on in Datapoint folklore in part of the story that Vic Poor told and retold of his astonishment when he found out what the buyers were doing with that first machine.

The most elaborate version has him flying to Arkansas in his private plane and getting a tour of the chicken farm, which culminated when they showed him the 2200. He asked what kind of telecommunications connection they were able to get, being in the middle of nowhere. (Noisy connections and even manual switchboards were still common in rural areas, making

---

68 The second installation was in Gainesville, Ga. The locations of the other two are not now known.
69 See http://www.generalmills.com/corporate/company/hist_SearchableBook.pdf, accessed August 7, 2009.

data transmission over phone lines via modems a doubtful proposition.) He reached behind the 2200 to find the telephone cord, could not find one, and was told they were using the post office for data transmission. They mailed the data tapes to headquarters, having written their own payroll program for the 2200.

Although he did later visit the chicken farm (and recalled that the installation he found was "not glamorous") his initial encounter was over the phone. He called a woman from Pillsbury who had attended his programming class and asked how things were going with the 2200s that Pillsbury had recently received. Hearing they were being used in a remote chicken farm, he asked what telecommunications protocol the machines were using, since he knew Pillsbury had not acquired any terminal emulation software. He was told that poor connections prevented the use of remote terminals at the chicken farms, so they had written a payroll program to run locally, on the 2200.

"We were pretty flummoxed over that—at least I was not anticipating it," recalled Poor.

Also, "I remember teaching our first customer class, in assembler, and it turned out that not a single customer was going to build a terminal," Poor said in an interview four decades later. "I had a whole course outline of how to program terminals, but the students kept interrupting, saying that we need to do this and that. I kept digressing to answer them, and screwed up my syllabus. That experience led to a conference with Harry, and we decided that we needed a high-level language."

The result was a programming language for CTC machines called DATABUS. Conceptually it was based on IBM Autocoder, a programming language for mainframes that had at least 4K of memory, but ended up not resembling Autocoder, Poor said. It was intended for business applications—there was no real use of graphics, in other words.[70] It is often compared to COBOL. Although machine-language programming was common in that era, after coming out with DATABUS, CTC never offered another class in machine-language programming for the 2200, Poor said.

CTC also quickly supplied an operating system based on tape storage, called CTOS. It let users create, store, and retrieve files from the cassette tapes. It required 6K to run. The previous stand-alone programs that CTC wrote were loaded and run by themselves using self-loader software at the beginning of the tape.

Recruitment of David Monroe

Gust's eagerness to show off the new 2200 was also instrumental in CTC's recruitment of David Monroe, who would have a pivotal role in the future of CTC/Datapoint.

As a freshman physics major at the University of Kansas, Monroe[71] had wanted to sign up for an introductory class in computer science, and found it was not open to freshmen. So he petitioned the head of the computer science department—Dr. Earl Schweppe—noting that he had studied computers on his own and had built a small one in high school.

---

70   DATABUS was a compiled language, and run-time software was eventually written for non-Datapoint machines. At this writing DATABUS lives on as PL/B, or Programming Language for Business, standardized as American National Standards Institute (ANSI) X3.238-1994. The name had to be changed, as ANSI cannot use trademarked names.
71   Monroe's recollections are based on an interview conducted in 2009, and follow-up e-mail.

He was turned down. Monroe turned to his physics advisor, who arranged for him to take a "special studies" class involving the use of computers in physics. Working alongside PhD candidates in a basement lab, Monroe managed to interface a Wang 700 electronic scientific calculator (a gift from his parents) to collect experimental data from a nuclear resonance spectrometer. Previously the data had been gathered on paper, and the area under the curve had been measured by actually cutting out that area of the graph paper and weighing it.

On the strength of that accomplishment, he applied for, and received, a $2 million grant from the National Science Foundation to computerize experiments in the physics and chemistry departments. This led to an organizational meeting that Dr. Schweppe also happened to attended.

"He kept staring at me, and after the meeting said, 'I know you from somewhere÷' I had the pleasure of telling him that I was the student that, about nine months prior, he said couldn't take Introduction to Computer Science. He invited me to his house for dinner that night and after that we became good friends," Monroe recalled.

In early 1971, when Gust was showing off one of the first 2200s in the Minneapolis area, he also traveled with it to Kansas to show it to Dr. Schweppe.

"He was having some kind of local computer show down there," Gust recalled. "We sat and played with it and I ended up trying to connect it to a telephone DAA.[72] If you did not get the wires right you blew up your communications box because there is power on the DAA wires. I did blow it up. One of the doctor's students came up and said, 'Let me look at that. Oh yeah, I can fix that.' That student was Dave Monroe."

Later, Dr. Schweppe contacted Vic Poor and suggested that he hire Monroe as an intern. Monroe, who had learned to fly as a teenager, flew down over the 1971 spring break in a Cessna, with his mother, his girlfriend, and his girlfriend's mother on board. He decided the place looked like fun, and started that summer.

"Little did I know that I was walking into the birthplace of 'desktop computing,' a.k.a. 'personal computing,' and networking," Monroe said.

During that summer, he encountered the TI chip that was supposed to emulate the 2200, as described in Chapter 6.

Curing Pain

The perception of the 2200 as a small computer was central to another early sale that has been recorded in detail, this time on the West Coast. At approximately the same time that Gust's first customer was making the chicken farm installation, future Datapoint sales executive Gerry Cullen[73] was following a commissioned CTC salesman around the Levi Strauss headquarters in San Francisco. As an engineer sent out from San Antonio to assist the salesman, Cullen was pushing around a prototype 2200 on a two-wheeled cart while the

---

[72] DAA or Data Access Arrangement was basically a direct-connect modem supplied by the phone company.
[73] Material in this section is from interviews with Cullen in 2008, and from Cullen's self-published book on industrial sales and marketing pitfalls, "The Coldest Call."

salesman sought the Levi Strauss executive who had the requisite "pain."

Cullen had been a sales engineer for DEC, first in Massachusetts and later in Houston. He made a sales call to CTC and ended up in a long conversation with Vic Poor and Jonathan Schmidt.

"We had a lot of interests in common, like circuit design and ham radio, and an interest in marketing," Cullen recalled. "I was interested in what people were going to do with computers. It was a big problem at DEC, as only scientists were using them and there were not that many scientists. On that or the second visit Jonathan chased me into the parking lot and asked if I would like to work there."

He started work at CTC in early 1971. "The sales pitch from Jonathan for the 2200 was that it was a programmable terminal," Cullen said. "They said they had no sales material and we'd have to write it ourselves. They said, 'Why don't you figure out what people think about this thing and do that for us?' That was my introduction to marketing. Vic Poor was very pragmatic and did not want to sell the technology, but the utility of the technology.

"The message we got, really quick, was that the customers already had a lot of terminals. They wanted to know if they could program the thing."

When Cullen was sent on the Levi Strauss sales call, he was unwise enough to ask why he was picked, and was told he was the least valuable person on the engineering staff, and was known to be patient with dumb questions.

In San Francisco, he learned that Levi Strauss was known to be considering buying scores, possibly hundreds, of small computers to put in the field for data entry. He accompanied the salesman, trundling the 2200 on its cart, as the salesman looked for the correct person to talk to.

Cullen quickly learned that talking to middle-level executives who seemed actually interested in the 2200, and who asked intelligent questions about it, was a waste of time. Those executives were yes-men who could perform endless analyses, but would not and could not make buying decisions. Instead, the CTC salesman had to find the decision-maker who had pain that the 2200 could cure. Typically, Cullen learned, this pain involved a trophy wife whose lifestyle had to be kept appropriately funded, since she could upgrade to a better-positioned husband in a matter of months. Her current husband, an executive whose personal finances she had over-extended, would be interested in championing a successful project that would lead to promotion and enhanced income, curing his personal pain. The task of the salesman was to identify that decision-maker, meet with him, and sell him the cure.

Finding that person took three weeks, during which Cullen turned on the 2200 exactly twice. The pained executive had an office with a view, with golf trophies and pictures of his tanned, beautiful brunette wife driving a BMW. He also had a tanned secretary who served coffee in china cups and saucers—no plastic or paper here.

There were several preliminary meetings where technical information was handed out. In the final meeting, Cullen was about to bring up technical specs, but the salesman waved him to be quiet. The salesman noted that Levi Strauss was an IBM user. Had they decided to not go with IBM on this deal? Or were they talking to CTC simply to get IBM to come down on their prices?

The executive said that IBM would not offer a discount, making their machine too expensive. Also, CTC's machine was smaller, which was desirable. But he was worried about teething problems with CTC's machine, since it was a new design.

The salesman suggested that Levi Strauss give CTC an order for four machines, payable in 90 days. If they were not satisfactory, CTC would take them back and tear up the order.

It was agreed, and the pain was on its way to being cured—with no mention of the 2200's performance or technical specifications. Levi Strauss would eventually acquire hundreds of the machines.

Cullen would spend the rest of his marketing career looking for pain points rather than product features.[74] But as for marketing the 2200, "There were two things the customer cared about: 'Can we program it ourselves?' and 'Can we send data over the phone lines?'", he recalled. "Transmitting data was a big deal. DEC did not do that very well. The business buyers had heard about IBM doing it, where it was really expensive. But we made it easy.

"I would ask customers why they liked it so much. They would say that they wanted to know how much each store sold at the end of the day, so they had someone key in the data and an evening report would go to the home office. They had been mailing it in, as this was before fax machines. We were cutting a week off their turnover time. Inventory, cash flow—everything was affected.

"People liked the looks of it, too. It was sexy. There was no big screen, no boxes, and no flashing lights or glowing tubes. It looked really friendly and did not look like it needed a scientist to run it. It looked like a fancy typewriter that anyone could use—like an office appliance. IBM's small machines were really the size of filing cabinets and took trained people to run. Our customers were training themselves to run the computers," Cullen said.

The Survey

What happened in the next few months was summarized by Bob McClure, the same professor who convinced Phil Ray of the need for plug compatibility. In late 1971 Ray called him again.

"He said, 'We have a problem—we started selling the 2200 and a lot of people bought them but then sales dried up. What's going on?' I said that I would try to find out," McClure said.

"I was able to do it because I knew what was going on in the computer industry. Some of CTC's customers were also (consulting) clients of mine. I said that I would call some users. I did, and found that they were very happy with the machines. But they were programmable, and they were busy programming them. They said that as soon as they got them

---

[74] The identification of customer pain points has since become a common theme in industrial sales training literature, by writers who have systematized the practices of successful salespeople. It is not sufficient that a proposed acquisition is a good idea or will save money—it must cure pain. However, the pain need not be caused by a spendthrift trophy wife, at least not in theory.

programmed and the applications deployed they would buy a whole bunch more. I talked to several people who said that the 2200 offered significant cost savings. The machines were not very fast, but were fine for data input.

"I had a long session with Phil and Gus with the results. I said, 'You have sold to the early adopters, who are interested in this class of machine, but they had to find if it would do what they wanted it to do in their environments. I think you will find that, pretty soon, your sales will pick up seriously.' And all those people who bought early machines turned out to be big users.

"They were all using it with in-house applications," McClure recalled. "Typically, they were looking to collect data in a smart way. The advantage is that they didn't have to tie up any mainframe time. That was a huge advantage because mainframe time was quite expensive. With the 2200 you could collect data, write it to a tape, and then dump it to a mainframe later on. A lot of 2200s were bought by order-entry firms or insurance companies. Previously they had punched the data in through Teletypes or through the half-duplex terminals that later became IBM 3270s."

Into the Future

During 1971, the CTC staff also began working on the next generation of the 2200, to be called the 2200 Version II. RAM chips had reached the market, and Version II would use those instead of shift registers. Consequently, the processor could access any memory address at any moment, instead of having to wait for an address to become available. That improvement, by itself, sped up throughput by a factor of anywhere between three and ten. The designers used more address space, allowing 16K of memory, permitting larger, more powerful programs. (Alternately, larger programs could run without the delay of loading "overlays" from tape.)

The Version II processor was a true eight-bit parallel unit, rather than a one-bit serial processor pretending to be an eight-bit unit. It included an interrupt line, which meant that a hard disk could be attached, allowing vastly greater and faster data access. However, the cassette tapes were retained for booting and software loading.

All in all, the Version II was often described as being a hundred times more powerful than the original 2200. The improvement was all the more dramatic because it could run all of the software that had been written for the original version, so any previous user could see the difference.

It would appear in the spring of 1972.

Meanwhile, design work had already begun on CTC's third-generation processor, to be called the 5500. It would have a faster clock speed, a capacious 64K address space for memory, additional machine-language instructions to speed up tasks, and segmented memory to allow multi-tasking. (Datapoint's processor dynasty is outlined in Chapter 17.)

It was while this work was underway in late 1971 that Intel came back with its one-chip implementation of the 2200 processor, then called the 1201 chip.

What happened next is probably best understood from a financial perspective rather than a technological perspective.

Chapter 8

CTC's Finances: 1970-1972

While CTC's engineers were struggling to create the first mass-produced personal desktop computer, during the same time frame its managers were struggling to, basically, keep the doors open.

As things turned out, it was not a good time to start a computer company, as 1970 and 1971 developed into the first recession ever experienced by the computer industry. When it began, IBM and its seven mainframe competitors were referred to as Snow White and the Seven Dwarfs, with wits proposing which company best represented Doc, Grumpy, Happy, Bashful, etc. Under pressure, RCA and GE (neither of which originally had an office equipment sales force) dropped out, leaving the BUNCH (Burroughs, Univac, NCR, Control Data, and Honeywell) and, of course, IBM.

And, as explained in Chapter 3, CTC was forced to rely on lease or rental agreements instead of outright sales. This meant that sales revenue from the products it manufactured dribbled in over a three-year period instead of arriving in one payment. But the expenses involved in manufacturing those products had to be paid up-front. So CTC was constantly short of money during this period, and constantly looking for new funds.

Given Roche's and Ray's previously described policy to favor time over money, the creation of the 2200 within an intense 100 days in the spring of 1970 can be seen as an effort to get the job done before the money ran out. And it did run out. They were able to raise more. That ran out. They were able to return to the well again, and finally saw daylight. It was a white-knuckle ride, and could have ended in bankruptcy and dissolution several times along the way.

As explained in Chapter 3, the startup money raised in 1968 was gone by the time of the initial public stock offering of September 1969. The IPO bankrolled the development of the 2200 during the spring of 1970, but that money was gone by summer—about the time that Intel stopped working on the 1201 chip. Meanwhile, Mazur was forced to announce that CTC was having problems collecting accounts from two customers, and the news caused the stock price to fall from $45 to $8.

So, in mid-1970, CTC's principals went back to the money markets to find more backing. The results were not pretty. In fact, Ray would later remember Monday, June 22, 1970, as one of the worst days of his life.[75] He and Roche flew into New York the previous day to present to potential investors and underwriters, hoping for a new round of financing. But before they could get started Monday the news broke that the Penn Central Railroad had gone bankrupt. The money men scattered, since handling their exposure to this disaster (the largest American corporate bankruptcy to that time) had first priority.

Roche and Ray were only able to get the ear of a man with an engineering background from New Court Securities, an arm of the Rothschild's that also backed startup Federal Express.

---

[75] No source actually gives the date, but it can be inferred from the evidence.

But in the short term nothing happened, and the burn rate continued.

As if things weren't bad enough, Poor remembered that investors were scared off by another market development that had nothing to do with CTC, the Penn Central, or even the on-going recession. Starting in about 1969, a company called Viatron Computer Systems Corp. of Bedford, Mass., raised a great deal of capital to sell a machine that sounded, superficially, like the Datapoint 2200. Designed for interactive environments, it had a 16-bit board-level processor, memory support for up to 16K (actually 8K of 16-bit bytes), two cassette tape drives for mass storage, a CRT and a keyboard.[76] Instead of a printer, the users were given a set of solenoids that they were supposed to set atop the keyboard of a Selectric typewriter, turning it into a crude printer.[77] But its rental price of $40 per month was about a quarter the rental price of the Datapoint 2200. How Viatron could afford to charge so little while using famously expensive magnetic core RAM was a mystery.

The apparent answer was that it could not afford to, and Viatron and went bankrupt in March 1971, leaving a legacy of lawsuits and wary investors. Worse yet, Viatron had used the term microprocessor to describe its CPU, tainting the word.

Meanwhile, CTC's 1970 financial year, which ran through July 1970,[78] showed revenue of $3,847,462, versus expenses of $5,064,019, for a loss of $1,216,557. This was on top of the 1969 fiscal year (CTC's first) when the firm had zero revenue and lost $688,000, which was basically every penny it raised that year.

Mazur would continue to make the rounds of venture capital and investment firms, but had no success. In terms of raising money, he was more successful dunning the two deadbeat customers, not balking at calling officials of the overdue firms at their homes, at their clubs, at their in-laws' homes, and, reportedly, even the home of a girlfriend of one of them.

As the summer of 1970 turned into fall, CTC began gearing up to manufacture the 2200. This evidently took longer than expected, probably due to the shortage of funds, causing turmoil among the sales force.

Enter TRW

In their search for additional money, Roche and Ray opened a new front, and began looking for companies that could buy CTC, that being preferable to financial collapse. They had done work with TRW Inc. when they were in the space program, knew people there, and knew TRW had the necessary funds. So they approached the defense/aerospace/automotive/credit-reporting conglomerate.[79]

For once, they got a nibble.

"TRW sent a team of four or five financial executives out to our company from the West Coast," Ray recalled. "They cut through the clutter pretty quickly. I admired them, having

---

76   See http://bitsavers.org/pdf/viatron/VCS-21-CR_010_Viatron_2140_2150_GeneralDescr.pdf.
77   See "A History of Modern Computing" by Paul E. Ceruzzi, page 253.
78   At the time, CTC was using a 52/53 week calendar under which the fiscal year did not always end on the same date, but for simplicity July 31 will be treated as the end of its fiscal year.
79   TRW was broken up after a hostile take-over by Northrop Grumman in 2002.

done satellite work with TRW. They didn't want to talk to the marketing people, they wanted to look at the product, and at the manufacturing costs, and see if it could be made for a lot cheaper than you could sell it for, and if people wanted it, and if it was something the world needed. They made a decision pretty quickly."

The decision was to buy CTC. The San Antonio startup would become a subsidiary of TRW, and its financial agony would end. Indeed, CTC's burn rate could probably have fit within a round-off error in the financial reports of the conglomerate, which ranked 68 in the Fortune 500. In 1970 it had sales of $1.585 billion.[80]

The recommendation was forwarded to TRW's board, which had a reputation for rubber-stamping such things. When the next TRW board of directors meeting came around with the CTC acquisition on the ballot, Ray and Mazur were in a nearby hotel room, in Los Angeles, waiting for the call saying that the deal had been approved and the documents were ready for their signatures.

But when the call came, they were told that the idea had been vetoed. No explanation was given, then or later.

"Their chairman went against the recommendation of his staff for the first time ever," Ray recalled.

Apparently to cover his embarrassment, the TRW vice-president they had been working with suggested that some deal might still be possible, at a smaller level of investment. Mazur had by then returned to New York for yet another futile attempt to raise funds, and was thinking that the end of the line might have been reached for CTC.

He was in his hotel room when the call came from TRW—and he had enough nerve to play hard-to-get. Yes, he lied, he was about to close various deals there in New York, but he could probably still find time to sit down with TRW again.

Mazur flew to California for the meeting. The deal TRW presented was, on the surface, stark and simple (although ironing out all the details took months.) TRW would participate in the next CTC stock offering, but in return TRW would get the exclusive right to market CTC's products overseas for five years. Additionally, TRW would get the exclusive right to manufacture CTC's products overseas for ten years.

That was asking a lot, frankly, but Roche, Ray, and Mazur saw little alternative. Not taking the deal could mean the end of CTC, and then overseas marketing and manufacturing rights wouldn't mean much.

So they agreed. Aside from having a conglomerate like TRW handle the overseas business (and this did prove advantageous) TRW's involvement gave CTC more credibility in New York. Consequently, other investors joined the next round of financing, including New Court. The first investments from TRW and New Court apparently arrived in October 1970, shortly before production began on the 2200.

CTC's interim financial report, issued January 10, 1971, noted "an atmosphere of confidence, in marked contrast to the uncertain environment prevalent at the beginning of last summer."

---

80 The figures are from coverage in the San Antonio Express News, July 11, 1971.

Winter turned into spring of 1971, which saw the first sales of the 2200 and customer acceptance of it as a general-purpose computer. But these sales, too, were leases. They showed that CTC had hopes for long-term survival, but did not solve CTC's fundamental economic dilemma. Development began on the 2200 Version II, but that involved still more up-front expenses.

The master agreement with TRW was inked in July of 1971. As a result of its association with TRW, CTC was eventually able to raise about $7 million.

Another financial crisis had been skated over.[81] There was little ground for celebration, as July also marked the end of CTC's 1971 fiscal year, and while revenue had actually gone down by a fifth, to $3,097,000, expenses had also gone up by about a fifth, to $6,847,000. So the 1971 loss that was triple the previous year's, at $3,750,000.

And now CTC had a new group of backers looking over its shoulder. The new group noted that neither Roche nor Ray had any real business experience, and Mazur's experience was limited to smaller ventures. So the new backers, led by New Court, sent out a man of broad experience in industry and banking, named Mike Faherty, initially just to look around and then report back to the investors. He arrived by the end of July 1971.

Mazur had been worn out by the last round of fund-raising, and disheartened by the death in March of stalwart CTC supporter Joe Frost Jr. Reportedly, he had clashed with Ray about Ray's spending decisions as president of CTC, and aspects of Ray's lifestyle. (There were rumors of pot smoking, which could have been disastrous to the company.)

Frassanito recalled that Mazur tried to fire Ray, but found that the corporate charter did not give him the power. Mazur's archives include an undated, handwritten note complaining about the board's failure to fire Ray at a previous, unspecified meeting, despite what Mazur thought was the members' previous resolve to do so.

Meanwhile, Mazur reportedly had disagreements with the newly arrived Faherty, who wanted CTC to continue spending money on product development. Mazur felt that such spending was unjustifiable.

The resolution came on September 1, 1971, when Mazur resigned as CTC's chairman of the board, ending his association with the firm he helped found.

As for Faherty, he apparently expected a cold shoulder in San Antonio, but found that the people at CTC were eager for help—for the most part. Faherty was an ex-Marine, and Frassanito recalled Faherty telling Ray that he wanted to run CTC like the Marine Corps. Ray responded that doing so would put the company out of business. (He also remembered Faherty questioning where the development money for the Datapoint 2200 went, while tripping over an actual Datapoint 2200.)

Faherty decided that CTC's finances were a mess. Its accounts payable was $1.2 million rather than the previously reported $600,000, and some bills had not been paid for six months. He also decided that its products gave it some prospect of success—if CTC could

---

81 Details of the TRW saga are from the Datapoint vanity history of 1982, the agenda of the 1970 stockholder's meeting, and Ray's final interview.

literally buy time to survive the leasing paradox.

It was during this period, when CTC's fate appeared undecided, that Intel brought out the 4004, and finally delivered the 1201 chip.

Faherty probably paid little attention to these developments. He reported back to the stockholders in December 1971, telling them there were two options. They could close their wallets and let CTC be consumed by its burn rate, and go out of business. That would probably lead to nasty stockholder lawsuits, so he did not advise it. The alternative was to come up with enough additional money to bring the Datapoint 2200 Version II to market. He expected it to sell well enough to save CTC, giving the investors some hope of "coming out whole."

The holiday season passed with no decision from the backers.

The CTC annual stockholders' meeting was held January 12, 1972. Faherty had two press releases written in advance. One said that a new line of credit would allow the company to maintain full operations. The other announced the commencement of bankruptcy proceedings.

Five minutes before stepping up to the podium, he learned that the first one was applicable—more money had been approved. But there was a condition: Faherty had to stay on as financial controller until the finances were cleared up—which they wanted done quickly.

And he would report directly to the board of directors, behind the backs of Ray and Roche.

That was quite a slap at Ray and Roche. They were the corporate president and vice president, respectively, and Faherty should have been reporting to them, not the board. Faherty did not accept the conditions until talking things over with Roche and Ray, who urged him to accept—they saw no other choice.

As for cleaning up the finances, Faherty took three steps. The first was a round of layoffs to cut expenses. Everyone was judged on their usefulness in the Datapoint 2200 II effort.

The second was to bring in a third-party leasing company to take over CTC's leases. The hardware that CTC had leased to its customers was sold by CTC to that firm, at a discount. CTC thereby finally got some cash up front for its products. The leasing firm then leased the hardware back to the customers at the original terms.[82] These were paper transactions, in other words. The hardware remained in place, untouched. The transactions eventually brought in more than $12 million. (They even sold, and then leased back, the back-office computers that CTC used internally.)

Subsequently, CTC would try to sell its hardware rather than lease or rent it, but if the customer wanted to lease or rent, CTC would use the third-party firm and get its money up front.

The third remedy was to impose a 2.5-to-one reverse stock split, causing the stock price

---

[82] CTC had had a third-party leasing deal with another firm in 1970, but apparently had lacked the clout to get favorable terms. Then and now, third-party leasing deals are a common arrangement with industrial equipment.

to rise from about $8 to about $20. Frassanito recalled. However, Faherty did it without consulting Roche and Ray, and did not happen to mention it until the four (Roche, Ray, Faherty and Frassanito) were in the coffee shop of Grand Central Station during another trip to New York. Roche was angry and showed it, Frassanito recalled, but by then there was nothing Roche could do about it. His mood did not improve when the effort proved futile—after rising as expected, the price eventually slumped back to the original $8 price, wiping out the gains, despite there being fewer shares in circulation.[83]

But from then on, CTC's story would no longer be one of white-knuckle suspense. There would be some borrowing at regular commercial terms, and some grumbling by the investors, until the turnaround came. Then the grumbling was soon forgotten.

As July 1972 came to the close, revenue for the 1972 fiscal year had nearly doubled over 1971, to $5,410,000, while losses had gone down by a third, to $2,220,000. Meanwhile, the recession had ended. All the indicators pointed to clearer sailing ahead.

But, unknown to anyone at the time, CTC's long-term fate had already been decided by a single decision taken during the previous autumn.

---

[83] Reverse stock splits are usually done to increase the price of a stock, to make it acceptable to institutional investors who often will not buy low-priced stocks. Reverse stock splits are not regulated by the SEC. Later, in 1980 and 1981, Datapoint would perform conventional stock splits to reduce the price of its stock and make it more attractive to small investors.

Chapter 9

"The Worst Business Decision in History"

The Intel 1201 chip was finally delivered to CTC in the autumn of 1971.[84] In hindsight, it's clear that this chip, through its direct descendents, was the foundation of the digital world. It was built to CTC's specifications and CTC owned the design. Owning its intellectual property (IP) could have represented for CTC what the first telephone patent represented for Alexander Graham Bell and his backers.

But, as detailed in the previous two chapters, CTC was then in dire financial straits. Also, CTC had not only done fine without the chip, but it had technology in its labs that made the chip look quaint. Consequently, in the short term the chip was worth nothing to CTC, and CTC would not pay for it.

"The chip was obsolete at that point, although not for some things," Poor recalled. "Intel said, 'Here is the part, we want our NRE (non-recurring expenses).' Phil said, 'You're a year late and we don't think we owe you anything.' Phil and Noyce met in San Antonio. They were good friends, so it was an amicable meeting. We agreed to waive any proprietary interest in the part and Intel could do what they wanted with it."

Some versions of the story say that CTC rejected the chip largely because it was too slow by orders of magnitude, as the MOS technology used by the chip was slower than the bipolar technology used by the MSI chips that composed the CTC board-level processor. Poor said that the real issue was that the chip was obsolete. He said the chip would have been about as fast as the Datapoint 2200 Version I, a machine that CTC was still selling but whose performance was about two orders of magnitude lower than that of the pending 2200 Version II. (In fact, he said that the 8008 could have been used in the 2200 Version I.)

Frassanito was not at the meeting when the decision was made. But he said that Gus Roche came to his office immediately afterwards, red-faced and shaking with anger. Frassanito understood that Roche had wanted to pay the money and retain the IP, but had been outvoted by Phil Ray and Vic Poor.

"Gus knew exactly what the chip was, and how important it was," Frassanito said. "Gus thought that it was an important breakthrough, and technically very elegant, and that it was important to own it.

"He was so pissed off—livid to the extreme—when they would not keep the intellectual property. He said, 'Those fellows don't understand this business and this technology.'

"Vic said, 'Why box yourself in? The customers don't care,'" Frassanito also recalled. "Frankly, in that era the cost of the computer was not in the chips, so using the microprocessor chip would not have affected the price of the product. So you can understand Vic's argument. But they gave Intel the keys to the kingdom."

---

84   No documentation for the date has surfaced, and the author has assigned a timeframe on the basis of various suppositions. But a few months either way would not have dramatically changed the circumstances.

He said the CTC even hired a lawyer to fend off any lawsuits from Intel to force CTC to pay for the chip. In effect, he complained, they hired a lawyer to ensure they did not end up owning the IP of the first microprocessor.

In his final interview, Ray did not mention any decision to reject the chip. Instead, he went back to his story that Intel undertook to develop the chip on a $100,000 bet with CTC that Intel could make the chip within a six-month deadline. By missing the deadline, according to Ray's story, Intel lost the bet, but was forgiven by CTC.

"Then the whole chip business got soft and the economy got bad and one day Noyce called and said, 'We can do this, but can we get off the hook? We have got to do some other things.' We had to remain friends with him—we had given his son a 3300, and he asked politely if they could table the project for a few months. So I said yes and the $100,000 never changed hands. We didn't ever dream that the computer on a chip would be that big a deal."

His story would be more credible if he did not then go on and say that the 8008 name came from the CTC purchase order number. Intel sources, of course, make it clear that this was not the case. Otherwise, his story could imply that the decision not to pay for the development of the chip, and the abandonment of its IP, could have been separate from the decision to reject its use. The best estimate is that Ray's six-month deadline (when Noyce would have wanted out of the bet) preceded the delivery of the chip by about a year.

The stories might be reconciled by referring to the recollections of Intel rep Bob McDowell. As explained in Chapter 5, he recalled that the development contract for the chip, between CTC and Intel, imposed a penalty of $50,000 on whichever party canceled the contract. This penalty was separate from the basic development fee. Perhaps it was this penalty that Noyce asked Ray to forgive. Then, when the chip was finally delivered, a bill was presented for the basic development fee, triggering the decision to reject the chip.

Results

Of course, it is easy to say that, from any business viewpoint, CTC's decision about the chip was fabulously short-sighted. Certainly, the end results are hard to argue with.

Its 2008 report[85] showed that Intel Corp. had revenue of $37.6 billion, and was the world's largest manufacturer of semiconductor chips. Of that total, $27.5 billion came from the sale of microprocessors. Basically, Intel spent part of the 1980s transforming itself from a memory (i.e., RAM chip) company into a microprocessor company, Japanese competition having made the memory market unprofitable. In 2008 it had only a handful of competitors in the microprocessor field, and its closest direct competitor (Advanced Micro Devices, or AMD) was about a sixth the size of Intel.

CTC/Datapoint, on the other hand, no longer existed by then. As the balance of this book will show, the decision to drop the microprocessor that it invented did not stop it from enjoying 10 years of phenomenal growth and success, as it found ways to leverage its original accomplishments. Then, the market changed, and Datapoint did not. Thereafter, its

---

85   The figures are from the Intel 2008 annual report, available at http://www.sec.gov/Archives/edgar/data/50863/000089161809000047/f50771e10vk.htm. Intel's fiscal year was also the calendar year.

story was one of disappointment, frustration, and decline, until its eventual dissolution.

Had CTC/Datapoint retained the intellectual property of the 8008, it would probably have initially charged royalties on the sale of the chip of about 5 percent.[86] At first, this would not have amounted to enough revenue to alter CTC/Datapoint's business model. After all, the initial market was tiny. Sales would have been higher for later chips, but the royalty rate would have doubtless fallen with each derivative generation of follow-on chips. However, its involvement would have given CTC/Datapoint more visibility and credibility. After all, its name would have been associated with the chip.

But after the market grew to a significant size, Intel would doubtless have developed an alternative chip that it could sell without having to pay royalties. That is what all of Intel's competitors did at some point. But even that move would have raised CTC/Datapoint's visibility, and the market for its chip would not have gone away immediately. Indeed, the enhanced competition might have brought about the personal computer revolution a little earlier. So if it did not end up becoming another Intel and dominate the microprocessor industry, it might have become a pillar of a more fragmented industry. Certainly, it would have been a different company with more diverse sources of revenue. The stock market debacle and consequent takeover probably would not have happened, and in the 1980s it could have faced the dramatic changes in its market without having been weakened by issues that had nothing to do with its product line.

Would that have allowed it to survive? Looking back, it seems fair to say that the more it would have become a chip company as opposed to a computer company, the better its chances would have been for long-term survival. But for CTC/Datapoint to completely transform itself into a chip company would have been asking a lot, especially as the microcomputer chip market was not born full-grown. Its prospects in 1972 were not as obvious as they appear in hindsight.

But hindsight also lets you dream. Even a one percent royalty on Intel's 2008 microprocessor revenues would have been about $275 million, which is more than seven times the profit that Datapoint rang up during its most profitable year (fiscal 1981.)[87]

There are many ifs, but one thing is certain: Texas' contribution to the PC revolution would be much harder to overlook, today, if the decision had gone the other way.

Carrying On

From a purely technological perspective, however, CTC's decision was not particularly short-sighted. The MSI chips that CTC/Datapoint used to create its board-level CPUs became increasingly sophisticated as the years went on. Consequently, the San Antonio designers were able to come out with increasingly sophisticated CPUs. Datapoint went for more than a decade without using a microprocessor chip in its main line of computers, finally making use of the 80286 chip in 1984. The fact that this move came just as the company began faltering was not the fault of any technology.

---

86 The average IP royalty in the electronics industry is supposed to be about 4.5 percent, but we'll assume that CTC had negotiating leverage, at least initially, due to the unique nature of the technology.
87 Inflation is not accounted for.

Unaware of the significance of what had happened, CTC went on trying to keep its customers happy with its new device—and was quite successful.

For instance, consider the case of Chuck Miller, who was made "technical project leader for mini-micro computer intelligent terminals" at the Johns-Manville conglomerate after a new CEO in 1971 decided to revamp the 19th Century way the Denver-based building materials company did accounting.

"I remember when the name Datapoint came up the first time, about 1971, and I thought 'intelligent terminals,'" Miller said. "We ran across glossies of the first Datapoint computer. Our management really put emphasis on it, as they wanted to move computing out to the plant managers. I was given the assignment.

"We reviewed every mini-micro and intelligent terminal on the market. IBM had the 5100 but the cost was outrageous and it did not do what we wanted. The only one that filled the bill was Datapoint. CTOS ran the same on all their machines. Managers could swap files between factories. No one else was doing that, instead they were selling one-off minicomputers.

"We started playing with Datapoint systems and we got some of the first 2200s with CTOS. They could do anything to anything, in terms of data collection. We soon had a guy trained on DATABUS. By the time we were finished we had more than a hundred machines. People loved them. The use of CTOS let us use the same training methods across the company.

"The term 'personal computer' was not used, as the concept had not arisen. We called them mini-microcomputers or intelligent terminals. They cost $10,000 to $50,000 each, depending on the peripherals. The concept of personal computer comes in when the cost gets lower. Datapoint introduced the flexibility to bring information technology to the people who actually needed it. They were cost-effective in what they could do and revolutionized the way we did business.

"Three things about it were wonderful for those days. First, the managers did not want to pay for it, but I was able to sell it on the cost of money for a line item, seven days versus overnight. Whatever their percentage of line item errors were, it would take seven days to resolve them. Previously, they would code all the accounts receivables, invoices, and shipments onto tapes and transmit then via a synchronous connection, with no error-checking, over noisy dial-up lines, so there were a lot of data errors. At the receiving end they would load the tapes (on the mainframe) and begin a billing cycle with data from all the factories. If they had an error they had to send a telegram back to the factory the next day telling them what they did wrong and how to fix it. Errors usually took seven days to resolve. But with Datapoint errors were resolved overnight. Each factory would pick up their error file the next morning and put it in that day's work.

"The second thing is that we had systems analysts touring the country teaching keypunch operators how to key order entries. They had 24 inches of ring binders covering how to code the sale of a product. The keypunch operators had to know all that in their heads, and it was a big burden for them. Of course they made errors. Where Datapoint came to bear is when we translated those 24 inches of binders into a DATABUS program so that the machine told the operator what information to get. It tied together all the things they were supposedly paid to understand but not paid enough to do so. There were five to eight analysts who were upset since they could not travel anymore—their movable party ended," Miller recalled.

The third attraction was the ability of the 2200s to serve as, yes, terminal emulators, and let managers perform on-line accounting and launch analysis programs on remote mainframes, he recalled.

"We trained product managers to use the datapoints, and the whole concept of remote computing was radicalized. Of course, once the novelty wore off they moved it to their secretaries, who were taught to push buttons and run standard reports," he said.

All in all, the corporate information cycle shrank from 45 days to one day, so that, for instance, managers knew the results of a fiscal quarter the day after it ended.

"The cost savings was enormous," Miller noted.

As for the technology, "The speed of the machines was more than enough for data entry," Miller recalled. "The later machines were much faster. I did not notice the speed difference myself but the programmers loved it. People understood that if they were doing something complex it would take some time, and they would start it and walk away.

"The time factor was in the time it took to send files. A 2400 baud modem was the top of the line and it could take 15 to 20 minutes to send the day's files. But they were satisfied with the way it worked." Programming a system to track bowling scores for the plant managers also helped with acceptance, he added.

Eventually they had Datapoint machines in scores of locations, he recalled. His staff would update programs by having their headquarters Datapoint machine dial into the remote Datapoint machines, transmit the updated code, restart the remote machines, and leave messages on their printers explaining what had happened and telling the operators to run appropriate tests.

"It always worked. It was bulletproof, since we used the same operating system at each site. Remote maintenance in 1973! We were doing things that are not often done even today," Miller said.

The only real maintenance issue that he remembered was that the power supplies of the Datapoint machine were sensitive to power spikes and static. If the operator was wearing nylons and crossed her legs, the resulting static would sometimes be enough to throw the machine off-line, so that the operator would have to restart it.

"We could use copper-impregnated grounded rugs, which were very expensive, or we could put fabric softener and water in a spray bottle and spray the rugs every day. That reduced the static," Miller explained.

Another Johns-Manville factory used a pressure stamp to make roofing shingles. The machine drew enough power to dim the lights. Stamping began every day at 4 pm, so they learned to turn off the Datapoint machine by 3:58 pm daily.

All the sites seemed to have a different phone company, but each site needed dedicated phone lines for the computers to enable remote operation. Local phone companies in the

Bell System invariably required 12 weeks' notice for installation. Datapoint delivered new machines in less than 12 weeks, which meant Miller could easily arrange for everything to be installed in one day. Other phone companies, however, could install a new line overnight.

As for CTC/Datapoint, "They were surprised by what we were doing. Since we were one of their first enterprise customers we had a great deal of input into how they did business. They would fly into Denver every six months, with technical people and vice presidents, and sit and chat for a day and get input. We would build a wish list for them, and nearly everything we asked for they did. It was always things intended to make it easier for non-IT people to run the machines. We were aiming it toward what we hoped would become personal computers. We were moving it toward those who were not technically trained, who just had keyboard skills and could read a screen.

"Datapoint fit a niche that had never been fit before. The reduced training, and error reduction was a quantum leap. They provided a plateau that was not there before. For its day it was leading edge," Miller said.

Chapter 10

Professionalization: 1972-73

After achieving financial stability in early 1972 (after unknowingly turning its back on the possibility of industrial domination in late 1971) CTC went through several important transformations during the rest of 1972 and 1973. Basically, it brought out the software system that would define its product strategy for years to come, it changed its name, it achieved profitability, and it started bringing in professional management.

By the end of the process, the company was firmly on the course that would take it through the next dozen years. During that time it would make additional contributions to technology, contributions that remain fundamental parts of the modern digital environment. But after that, it would find that its past accomplishments carried less and less weight.

New Market, New Name

During this period CTC brought out the Datapoint 2200 Version II. As previously explained, it was as much as 100 times more powerful than the first version, thanks largely to the use of RAM memory and an eight-bit data path. CTC also started offering optional hard drive storage for the 2200, and a disk operating system. In practice the Version II was always sold with a hard drive, in a separate enclosure. The cassette tape drives on the top of the 2200 were retained for software installation.

During this period CTC also brought out DATASHARE, basically a multi-user version of its DATABUS business-oriented programming language. Instead of one person running one DATABUS program while sitting at a 2200, multiple 3300 terminals could be attached to the 2200, each simultaneously running a DATABUS program on the 2200. A given terminal might be located in the same room, or located remotely, operating through a telephone modem. The applications were usually straightforward, such as data entry. The DATABUS software would check the data for validity, consolidate it and send it, error-free, to the corporate mainframe. The number of terminals that could be supported varied according to the complexity of the application, and expanded over the years as the processors got faster, but was rarely more than a dozen.

Multi-terminal DATASHARE systems remained the bread and butter of the firm's product line for the next decade. They were typically used as front-end communications processors for corporate mainframes, feeding them the data needed for general ledger, payroll, inventory, and similar business accounting applications. In some cases these accounting applications could just as easily been performed by the Datapoint processors themselves, but the trend of moving mainframe applications to desktop computers did not emerge for more than a decade. Unfortunately, when it emerged, that trend would rely on generic PCs, not Datapoint machines.

DATASHARE and the 2200 Version II depended on each other, since processor throughput became an issue with DATASHARE.[88] With the simple, single-user data input programs that were standard on the original 2200, the slow processor speed was adequate since it was still faster than the reaction time of the user. But with more than one user on a machine, this was no longer the case. CTC users now had to contend with response times, an issue that had been intensely studied in the minicomputer world for the previous decade. Typically, the response time rose incrementally as each new user was added, until it hit what was called "the knee," at which point it leapt toward infinity.

The firm also came out with software called SCRIBE, turning the 2200 into a basic word processor.

All CTC software, incidentally, was free to all hardware buyers, who merely had to pay for the cost of the transfer medium (such as the tapes.)

As for the corporation's name, CTC's managers were finding it to be a hindrance. They were now selling minicomputer systems based on the Datapoint 2200, and so the Computer Terminal Corporation was no longer a computer terminal corporation.

The issue supposedly came to a head when Roche and Ray arrived at the Connecticut headquarters of Xerox Corp. for a scheduled meeting, and the receptionist had not heard of them.

There was a Datapoint 3300 terminal sitting on her desk. "We make these," Ray told her. "Xerox has hundreds of them."

She then remembered that there was indeed an appointment for "Datapoint" and they were ushered in.

Frassanito also recalled seeing brand recognition surveys in which respondents recognized CTC and Datapoint as separate brands. IBM, he recalled, got first-place recognition in all categories, even categories in which it had no products.

Going with the flow, so to speak, the company's name was officially changed to Datapoint Corp. during its board meeting of December 7, 1972.

Along with a name, the firm also adopted an identity during this period, based on the market it was serving. Gerald Cullen, who was involved in the early Datapoint 2200 sale to Levi Strauss, was by this time fully involved with marketing. He recalled bringing in an outside ad agency to shape an advertising campaign for Datapoint.

The advertising executive met with Datapoint's management, and Cullen remembers him turning to Phil Ray and asking what business Datapoint was in.

"Phil was one of the most articulate people I have ever met, and he said, 'You know, those mainframes are really big, and it seems like people want some computing power out in the field where they are,'" Cullen remembered. "The advertising man said, "So, you mean distributed computers?' Phil said that IBM had distributed computers. The ad man said, 'So you must be in dispersed computing. Fine, you're the leader in dispersed data processing.'

---

88 David Monroe pointed out that the firm's new reliance on the DATASHARE market was a major reason for turning down the 8008 chip, since, being comparable in power to the 2200 Version I, it could not run DATASHARE.

"It changed everything—we had an identity then, and a whole new motto. We admitted we sold little computers—up until then we still had a programmable terminal flavor. And we were the leader of something," Cullen said.

The resulting motto, "Datapoint, The Leader in the Field of Dispersed Data Processing," would be a recurring element in Datapoint corporate literature for at least the next decade.

The Bahamian Episode

Meanwhile, the spirit of improvisation lived on in the company's operations, sometimes with uncomfortable results. For instance, during this period Datapoint's management decided to hold an international sales conference in the Bahamas. They wanted to show off some of the new Datapoint products there, but discovered that no commercial airline had a container service that connected there.

As it turns out, there is a network of freelance pilots with small planes that serve the funeral industry, delivering encoffined deceased persons to any place served by any kind of airport. They located such a pilot in San Antonio, whose Cessna 182 had doors modified to accept a coffin, making it suitable for computer and printer consoles.

David Monroe accompanied the shipment, and soon saw that the pilot was not accustomed to accommodating passengers who were still alive. He chewed tobacco and calmly spit on the floor of the plane. Monroe, himself a pilot, was in the habit of flying around bad weather. This pilot thought it was great fun to fly straight through, and experience the violent, lurching updrafts, downdrafts, and lightning.

They made it to Miami, where they rented over-water navigation gear. They then made it to the Bahamas—where Monroe's problems were just beginning, as he was detained by Bahamian customs, demanding a $30,000 bond for importing the computer equipment into the country.

Monroe did not have $30,000. Nothing loathe, he signed the plane over to the Bahamian government as surety, not bothering to tell the pilot-owner that he'd done so. After the conference, on the way out of the country with the reloaded computer equipment, he was able to reverse the bond. They made it back to Miami, where Monroe decided the shipment no longer needed his attention. He returned to San Antonio in a first-class commercial airline seat, luxuriating in its comfort and cleanliness.

Perhaps foreseeably, the pilot of the flying hearse was killed in another plane a few months later flying through bad weather in Latin America.

"I really felt lucky then," Monroe recalled.

Into the Black

Fiscal 1972 had been uniformly unprofitable, but the loss had been progressively smaller from quarter to quarter. A loss of $626,000 in the first quarter fell to $600,000 in the second quarter, $561,000 in the third quarter, and $433,000 in the fourth quarter, which ended at the end of July 1972.

In the August 15, 1972, issue of the Wall Street Journal, the firm's earliest known big-sale press release appeared, stating that the Canadian National Railway Co. had purchased, through TRW, 149 "Datapoint terminal systems" worth $2.9 million.

After the first quarter of fiscal 1973 (which ended with October 1972), the company showed a profit for the first time, of $156,000. The trend continued into the second quarter, with a profit of $389,000, and on through the rest of the fiscal year.

Just before Christmas of 1972, Phil Ray did another press interview, and said that the firm (now renamed Datapoint) had weathered the downturn. They were now selling 160 Datapoint 2200 systems per month. Besides the Canadian railroad, major users included an unnamed West Coast drugstore chain that was using the Datapoint 2200 in 60 stores. Datapoint now had about 335 employees. It was represented in 16 countries (thanks to TRW) and overseas orders amounted to nearly a third of its backlog.

"We've spent four years developing this market and right now we've got it pretty well to ourselves," Ray said.[89]

In the spring of 1973, Datapoint's South African distributor announced three major orders worth about $2.5 million.

When the fiscal year ended the next July, it would show a profit of $1,957,000, and revenue that had increased 245 percent from the previous year, to $18,645,000.

At the end of October 1973, Datapoint would boast of having sold, to then, 3,071 Datapoint 2200s, to a total of 409 customers. The retail price of the machines ranged from $6,040 to $13,297. The monthly lease price ranged from $167 to $360 per month.

The investors slowly realized that they could breathe easily. Indeed, Datapoint would remain in the black for the next 12 years.

But for the founders, the black ink was a different kind of symbol—that it was time to move on. Now that the firm was on a sound footing, they began looking for an experienced industry professional to run it, and replace themselves.

"It's like designing and building a new airplane—to me it's fun and exciting, while being an airline pilot would be boring as hell," Ray said in his last interview. "Gus and I had a target, and we were determined not to bring in anyone until this thing was in the black."

The official story, from the unpublished 1982 Datapoint history, is that they assembled a list of candidates from various sources. Most were quickly eliminated for various reasons. One that was not was Harold E. O'Kelley.

Originally an electrical engineering teacher at Auburn University, O'Kelley had left that job to become a project engineer at Radiation Inc., which made (among other things) transmission equipment used in the space program. He rose through the ranks and by 1964 was in charge of all government contracts for the firm. The government division was already the largest in the company, and in four years he doubled its size. Radiation merged with Harris Intertype in 1967, and O'Kelley rose to be head of Harris' broadcast and composition equipment

---

[89] San Antonio Express-News, December 24, 1972.

group. Harris was a $300 million electronics conglomerate headquartered in Melbourne, Fla. O'Kelley's group accounted for $80 million of its revenue. By comparison, CTC's revenue in the previous fiscal year had been about $5 million.

According to the story, Roche made the first contact. O'Kelley said no, but was polite enough to agree to look at the corporate literature that Roche offered to send. The financial information turned him off—but he was intrigued by some software documentation that came with it. Solely on the basis of that, he came to San Antonio for an interview, and saw that the firm had made a profit of $156,000 for the latest fiscal quarter. (In other words, he must have come after October, 1972.) He was intrigued by the products. He said he'd think about it. He went home.

Months went by without a decision. Roche kept calling him and sending him information.

Finally, O'Kelley told his superiors he had gotten an offer. They made a counter-offer that CTC could not have matched. O'Kelley asked Harris' chairman of the board about his prospects if he accepted Harris' counter-offer. He was told that, at Harris, he would certainly become a millionaire—eventually. If he left Harris for CTC, he had a remote chance of becoming a multi-millionaire.

Against that was the realization that he was 48 (near the traditional age for a mid-life crisis) and it was clear that there was no chance of being promoted at Harris for six or seven years. If CTC did not work out, there would be other start-ups.

So he took the plunge, becoming Datapoint's president on March 11, 1973, about six months after the initial contact. Phil Ray, previously the president, was eased upstairs to become the chairman of the executive committee of the board of directors.

Frassanito remembered that the circumstances of O'Kelley's recruitment were less auspicious, and other sources give a similar story. "Phil and Gus set about looking for a replacement who would understand that business, and called friends at Harris to ask about who might be available. One of the people they talked to was O'Kelley, who said that he himself might be the right guy. So he flew to San Antonio to be interviewed.

"I interviewed him in the conference room, and saw that he was uneasy about being there since no one (at Harris) knew that he was interviewing," Frassanito.

O'Kelley was right to be nervous since another ex-Harris employee in the building noticed him and called a friend back at Harris to find out what was going on. As a result, word reached O'Kelley's boss, who asked him if he had been looking for a job. O'Kelley reflexively denied it, possibly because he had decided to say no to Roche's offer, and walked out of the room. Then he realized that his boss must have known what was going on to have asked the question, and went back inside and confessed. Having made a bad impression with his boss, he decided that he'd better take the Datapoint job.

New Sheriff

O'Kelley immediately began preaching growth—serious growth. For instance, he immediately announced that the company would remain a single profit center until it surpassed annual revenue of $100 million. Then he would consider delegating. Yet, when he arrived, Datapoint had revenues of only about $1 million monthly.

He would continue preaching growth, relentlessly, for the next nine years—until the resulting growth obsession abruptly triggered a catastrophe. But in the meantime the results were hard to argue with, as Datapoint's annual revenue growth from fiscal 1974 through fiscal 1981 averaged a stellar 49.5 percent. Business magazines of all stripes began seeking interviews with O'Kelley, as he seemed to have the Midas touch.

From the start, he was clearly not trying to win a popularity contest. As a result, even compliments about his style tend to be backhanded. "He was able to resolve the petty differences and resentments that had developed among the executives because of the financial problems," Frassanito recalled. "He terrified everyone and they had to drop their petty issues to survive."

Frassanito would later count O'Kelley as one of his friends. But when O'Kelley arrived at Datapoint, Frassanito got the impression that O'Kelley had decided to immediately make an example of some executive, and had settled on Frassanito. After all, he was young, an artist, and in a staff rather than a line position.

Frassanito had ordered a sign to be installed on a Datapoint building, but it was delivered two days before he returned from a vacation in Mexico.

"O'Kelley got mad and called me and said I was supposed to be there—and that in two weeks we can have a meeting about it," he recalled.

"Two weeks go by and I go into the meeting. I thought it was a trivial thing since I had only been gone for two days. He proceeded to yell and scream. His face turned red and the veins in his neck bulged. 'If I think you have a problem then you have a problem,' he yelled. He ranted and raged about responsibility.

"I walked out in shock—I had not been yelled at like that since I was in the army. Gus and Phil were in the next office. I said, 'Gus, this is crazy. I won't stand for this.' Gus said, 'Did he fire you?' As a matter of fact, I didn't recall that he had. Gus said, 'Well you know he could have if he had wanted to, and if he didn't fire you with all that yelling he's not going to.'"

But proactively firing people was not O'Kelley's style, it developed. "O'Kelley could be very cold," recalled Ed Gistaro, whom O'Kelley later brought over from Harris. "I was at a meeting with O'Kelley (while both were still at Harris) and they were having a manufacturing problem. They brought in the head of manufacturing and the poor guy was scared to death. O'Kelley drew his information on the blackboard and said, 'What you are telling me is that you did this, and this, and this.' And then he drew a circle around the information, and said, 'This is your job description and this (the circle, as a zero) is what you're worth.' That guy was gone before O'Kelley was out the door. Another time he threw me out of my own staff meeting. He was a very gruff guy, and hard to deal with, but I loved the guy."

"He would pick out one person at the conference table and denigrate and embarrass him in front of his peers, until he quit," Frassanito recalled. "He never fired anyone, but he tortured them until they quit."

"There was no word of complaint about management until O'Kelley arrived," Gerald Cullen recalled, with a viewpoint that was lower on the totem pole. "Many strange guys were hired after that. Phil Ray, Gus Roche, and Vic Poor were all about the products and the customers. O'Kelley was all about administration, throughput, and performance. He was an autocrat, and had no passion for the products. You could always go to Phil and complain, but O'Kelley immediately formed an inner circle, with gatekeepers. It was a culture change, but we got through it."

But, decades later, two things about him nearly always emerged from the memories of those who knew him then.

The first was his insistence that the corporate logo be changed because it pointed in the wrong direction. The old CTC logo was a stylized C whose interior was an arrow pointing to the left. The right is the direction of progress, O'Kelley insisted. The counter-example of NASCAR racetracks leaped to the minds of many employees who were Southerners, but they kept it to themselves, apparently. A new logo was introduced, and retained for the rest of Datapoint's history, using a stylized D whose interior was a triangle. The triangle pointed to the right, of course

The other was the impression they retained (fairly or not) from some of the fawning magazine interviews, that O'Kelley had supposedly arrived to cleanse the Datapoint financial temple, so to speak. The prodigal company had been adrift in a sea of red ink, until the sagacious O'Kelley rescued it, etc. As has been shown, Datapoint was in the black before he arrived, and O'Kelley was recruited only because it was in the black.

"I don't think O'Kelley deliberately said anything inaccurate, I think he was misquoted," Ray said in his final interview. "It irks me to see stuff like that."

What O'Kelley genuinely accomplished on arrival was to revamp Datapoint's marketing deal with TRW. As explained in Chapter 8, two years earlier the firm had signed over its overseas marketing and manufacturing rights to TRW in return for investments from the defense conglomerate. Soon after his arrival at Datapoint, O'Kelley set out to raise working capital by either launching another stock offering, or by selling convertible debentures.

The underwriters balked at doing either—or having anything to do with Datapoint. They pointed to various vaguely worded clauses in the TRW agreement, which they said allowed TRW to manufacture any Datapoint product overseas, and then import it into the U.S. and undercut Datapoint. Meanwhile, Datapoint was not allowed to perform any overseas manufacturing. Datapoint didn't agree, and anyway said that the overseas manufacturing rights only covered products that Datapoint was making at the time of the agreement (i.e., the 2200, the 3300, and some peripherals.)

It was an impossible situation, and O'Kelley later said he would not have taken the job as Datapoint CEO if he had understood it in advance.

He commenced lobbying TRW to change the terms, but they didn't prove very motivated to negotiate. (On the other hand, they evidently weren't doing any of the feared importing.) After three months he got their attention by offering to stretch their overseas marketing rights from five years to ten years, if they would drop the overseas manufacturing rights.

But TRW would have to actively market the Datapoint products, rather than just accept commissions. The revised agreement was inked in July 1973. The next month he was able to sell convertible debentures[90] worth $8 million.

Then, he set out to professionalize Datapoint's sales department. That chore was largely handled by another man he brought over from Harris, Ed Gistaro.

New Deputy

Gistaro had been involved in various marketing functions at Harris, but always dreamed of being the marketing director of a $20 million company.

"I really liked product marketing, where I was working with guys who are trying to bring products to market, explain them to the public, get people to buy them, manage the people who advertise them, and train the people who are going to sell them, etc.," he said. "It looked like Datapoint was perfect because no one had done that in a professional way there." He accepted O'Kelley's offer in the summer of 1973.

His first assignment was to turn Datapoint into a $100 million company in five years,[91] by building up its sales force. Previous retrenchments had reduced the sales force to only about 16 people. Sales management had been chaotic.

Gistaro recalled that, "The vice president of sales at that time came up to me—he was a really gruff old Texan—and said, 'I want you to know two things that I don't like to do. I don't like to meet customers, and I don't like to travel.' And he was the vice president of sales! It was so funny. I nodded, and got rid of him eventually.

"One of the first things I did was set up an industry-startling compensation plan. I told the salesman they would make six percent commission[92] straight up with no caps. The financial people went crazy, saying guys could make a lot of money. They could make $100,000! I said that they could make a lot more than that. In fact, later, we had a salesman in New York who sold the Citibank account $10 million in one year and made $600,000. That was far more money than anyone else in the company was making, including the CEO. But O'Kelley said we made a lot of money off that $10 million, let's not begrudge the salesman.

"Our compensation plan was way more lucrative for top-performing salesmen than anything IBM or Data General could offer. They had caps, they had much lower commission rates, and they had declining scales so that you made less on your second million then on your first million. My argument was, why is the third million worth less to the company than the first million? It didn't make any sense to me—when you have one guy selling as much as three, everything is cheaper in terms of office space and other things, except for the compensation plan," Gistaro said.

Meanwhile, the Datapoint sales force needed every possible incentive and motivation, since they were in an eternal, uphill war against a well-entrenched IBM.

---

90 A convertible debenture is a low-yield corporate bond that can be converted to stock at some point in the future.
91 In fact, it only took three years.
92 Commissions actually varied according to the product and on-going marketing promotions. Six percent may be considered an average.

"IBM could afford to assign a salesman to just stay at an account," Gistaro explained. "The account (i.e., customer) would give him an office. A lot of the corporate computer managers in that era had previously worked for IBM[93] and were effectively still on the IBM payroll. They were getting paid by the company they worked for, but they were doing what IBM told them to do. As long as things went okay they had that IBM security blanket. No one ever got fired for buying IBM equipment. IBM made him what he was, and was keeping him in his job, so trying to change that guy from an IBM guy into a Datapoint guy was really difficult. When selling a system instead of just one machine, you had to go through many more layers of management to make a sale, and at each layer threaten their comfort with IBM. We were pretty successful considering we had to fight that.

"We beat IBM on functionality and price. One of the big concerns you have when you make a commitment to a vendor is post-sales service. What happens if it breaks? If the guy says that they don't handle that, then you're kind of skeptical, and you're really skeptical if it's a computer system. But we had a good service organization. In fact, it was unusual for a company our size to have one.

"The salesmen that joined us were pretty much the pioneers in the industry. They were the guys who first put in desktop computing and first put in networking. They were the guys who stuck their necks out, and they did very well because of that, because it was the beginning of the revolution that we are still seeing," he said.

Datapoint would eventually have a sales force of several hundred, but flawed sales management would later play a major role in the company's downfall.

All Dressed Up...

Perhaps inevitably, O'Kelley also decided that Vic Poor was the technical expert he would rely on. That tended to sideline Gus Roche and Jack Frassanito. (Phil Ray had already been sidelined into the board of directors.)

"This is not working out, we decided," recalled Frassanito. "We can stay here, and no one will fire us. Or we can start another firm and do it again."

They began searching for another technology business to start. It could consume them for the next two years—Roche literally for the rest of his life.

Also during 1973, Ray and Frassanito took a newly hired marketing executive to dinner to welcome him to Datapoint. The newbie excitedly announced his intention to immediately buy a house and move his family to San Antonio, and thus demonstrate his commitment to his new employer and to the community.

Ray and Frassanito exchanged glances. They had been in town for five years and had not even considered buying houses. They had turned Datapoint into a going concern, but mentally they were still in start-up mode.

So they began the process anew with another firm.

---

[93] The Brotherhood of Men in White Shirts is discussed in Chapter 3.

Chapter 11

Gus Roche's Second Act, and Final Act: 1974-75

Sources agree that, after O'Kelley's arrival in March 1973, Phil Ray's health turned bad, and he had to have an operation. He fully recovered, however, and together Roche and Ray began looking for other projects.

After about a year of research, and of writing and rewriting business plans, they settled on a plan: they would develop memory devices around charge-coupled devices (CCDs). They were invented in 1969, and won their inventors the Nobel Prize in 2009. Since they can receive charges via a photoelectric effect, CCDs are now used as sensors in digital cameras. CCDs were much slower than contemporary RAM chips, but could be much denser. Their 45-page business plan (written by Frassanito) noted that RAM chips then on the market, made with the MOS technology, offered 4K bits, but CCD devices could offer 16K or even 32K bits. In fact, they anticipated jumping to 64K and 128K. To maximize production yields, they would use an adaptive control system to work around faulty memory segments.

They would call the company Mnemonics Inc. The initial product they envisioned would be a solid-state removable cartridge holding 2.4 megabytes. It would have been plug-compatible with contemporary removable cartridge disk drives, which were often used to be transfer software in that era. However, the Mnemonics product would have been about 140 times faster. (It still would have been at least a thousand times slower than RAM, however.) Meanwhile, they anticipated taking advantage of the relentless pace of improvements in the power of semiconductors, allowing continually faster devices, with more memory, at the same or lower prices. Mechanical disk drives would not be able to compete with that pace.

In completeness, the plan was in a different league from the original CTC plan, laying out every dollar that would be needed for the first three years of operation, including equipment rental, and the cost of fitting out the factory shell they hoped to have built in northeast San Antonio, where land was cheap and plentiful. (Datapoint's facilities were grouped in the more developed northwest side.) There were only about 35,000 industrial workers in San Antonio, but Datapoint's experience indicated that the area possessed an abundance of "easily trained direct labor and was attractive to professional employees recruited from other areas."

They decided they needed $3,300,000—and were able to raise it with little difficulty. (However, subsequent results showed that they would have been wiser to follow that policy of Gerald Mazur when the raised the initial money for CTC—double the estimate and add $50,000.) Roche, Ray, and Frassanito left Datapoint in July 1974 to run Mnemonics, as their lawyers told them that they couldn't remain officers of one company while starting another. They set up facilities in both San Antonio and in the area south of San Francisco that is now called Silicon Valley, and work began in earnest. Datapoint invested $50,000.

In a reprise of what happened in 1968, Roche assured Frassanito that the CCD memory device was just a way to get the new company started, akin to CTC's original 3300 glass Teletype. The equivalent of the 2200 (i.e., the eventual goal of the startup) would be wafer-level integration, which might eventually lead to a super-computer on a desktop. Then (and now) the microscopic circuitry that makes up chips are printed on precisely made disk-shaped silicon wafers with the circuitry of each chip forming a tiny square called a die. After the circuits are laid down the wafer is cut into is component dies. The dies are then tested and the bad ones discarded. The good ones are then mounted in plastic housings with connector pins to make the chips. Roche's eventual intention was to build multiple kinds of dies onto a wafer, including processors and memory, but then leave the wafer intact. Intelligence would be built into some of the circuits to test the dies and block out the bad ones. The result would be a powerful computer with multiple processors, Frassanito recalled.

Also involved in the start-up was David Monroe, Carver Mead, and Amr Mohsen. Mead was a Caltech professor who went on to receive the National Medal of Technology from President Bush in 2002. Mohsen was an Egyptian-American with a computer science doctorate from Caltech.[94]

Monroe soon got sick of flying back and forth between Texas and Silicon Valley. Using facilities supplied by Frassanito (who also provided industrial design and experience with human factors) he and Frassanito began tinkering with a computer-controlled two-way closed-circuit TV system to achieve what is now called videoconferencing. Because of a convoluted series of events, eleven years later it would have enormous impact on the fate of Datapoint. And, 13 years after that, Datapoint would have an enormous impact on the inventors (as described in Appendix A.)

This time, Roche told friends, he planned to remain in control. Some got the impression that he hoped to be successful enough to buy back Datapoint.

Whatever else the founding of Mnemonics meant, it also meant that there were now two companies that bore the unique stamp of Gus Roche.

The Roche Touch

"Father had the philosophy that if he spent tax-deductible money to give his engineers space in the plant and let them do their own projects, they would complete their corporate projects with more vigor and they would keep their creative sides alive," explained Chris Roche. "Plus, they would be there on the weekends. The key people behind the 2200 loved being there."

Roche was also typically there on the weekends, often bringing his children with him. "We would drive a go-cart in the parking lot, or play with a Selectric, or play with the oscilloscope. After they moved to the big plant (in northwest San Antonio) we would go downstairs to the sheet metal shop and make things," said Chris Roche.

"He could go in at least on Saturday, because the engineers and development guys were always working on Saturdays, and often on Sundays, because they found it so compelling and interesting," said Austin Roche. He recalled his father bringing him more often after CTC moved to the bigger building.

---

94 In 2007 he was sentenced to 17 years in federal prison for perjury, obstruction of justice, and witness tampering for acts related to an alleged murder-for-hire plot that originated with a patent dispute not connected to Datapoint or Mnemonics.

"I just got to hang around because my father would be busy. I would stand by while he talked to the guys about their designs, or what was happening on the factory floor. He might set me up with a spare oscilloscope. Or he would put me at the soldering iron bench, throw down some components, and say, 'Play!' I would put some things together and put probes in and mess around with the knobs."

He recalled watching them grapple with the question of how to "burn-in" Datapoint 3300 terminals prior to shipment, and for that matter, how to package them for shipping. The power supply could not be gotten to work immediately, he remembered.

"I had a distinct feeling that there were no rules, and not much protocol," Austin Roche remembered. "Everything was based on merit and accomplishment. You could tell it was a dynamic, innovative environment." People would work long hours for extended periods, burn out, and stop coming in until they recovered. Regular hours came after Vic Poor arrived, and the design and manufacturing engineers were separated into two camps, he added.

"No rules" summed up the experience of David F. Reed, who later became a systems architect for processor vendor AMD. Born in San Antonio but raised in Mexico, in 1973 he was attending a class in modal logic (a branch of philosophy, having nothing to do with binary circuits) at San Antonio's Trinity University. He mentioned to an older woman in the class the he was looking for a job. He also mentioned that he was a ham radio amateur. She said her husband was a ham, too, and sent Reed to talk to him.

Her husband turned out to be Vic Poor, who passed Reed on to a manager who gave him a job as a technician at Datapoint. Despite having a major in philosophy, with a minor in languages, he found himself wire-wrapping and debugging prototype circuit boards.

"They were not what I could call prejudiced about degrees," Reed said. "It was a great place to work because they had too much to do. When you wanted to take on something, you were likely to be given a free hand. As you did more they became more confident in you and promoted you.

"I was invariably working long hours, but since I felt like I was contributing and getting stuff done I felt no resentment. It was an opportunity to do more and get recognized for it. I learned that you don't get ahead without producing something, that you don't produce anything without hard work, but if you do produce people treat you fairly. That was the general environment. I know that I was not the only one who got ahead with a lack of degrees," Reed said.

Joel Norvell had started working at Datapoint in 1974, doing software development, disk diagnostics, and data communications. Working conditions had deteriorated for Datapoint technicians by that time, he said, with all activity taking place in one noisy "bullpen." But Mnemonics operations were taking place in a neighboring building, using borrowed Datapoint equipment. Working in the evenings for relief, he met Gus Roche, who often came by after hours.

"He was just talking to people and hanging out," Norvell recalled. "He was very excited about the next firm (Mnemonics). We talked about Greek culture and his favorite movie. He was the first visionary I ever ran into. He saw where things were going and was extrapolating it into the future.

"He tended to elevate the level of discourse in a positive way. The two people I encountered that reminded me of him were (Intel cofounder) Bob Noyce, who I did chat with at Intel, and (inventor and futurist) Ray Kurzweil, who I had heard him talk a couple of times. They had the same attitude and method of operation.

"Gus was a business guy in a rumpled suit, with a tie that was loose by the end of the day. He would talk about his vision and share it. He had lost control of Datapoint but he thought Mnemonics would be another home run, and he thought he knew better how to remain in control this time. Maintaining control of the company would let him realize his vision.

"He put effort into getting to know people. It was not just extroverted behavior but a factor of his being—he depended on other people to bring him perspectives. Knowing him was a moment of glory for me in a certain sense. He gave purpose to things in some intangible way," Norvell said.

Norvell also said he talked to Roche about the latter's intentions when he and Ray founded CTC. "He said that when it came to raising money it was more of a slam-dunk to describe the terminal product rather than describe the evolutionary product that was within the terminal. But creating a microcomputer was the original intention—it was not an afterthought at all. I had that conversation with him in 1974."

Nerd Heaven

The end result of Roche's management attitudes was something the participants referred to, without embarrassment, as "nerd heaven."

The summer of 1974 when Austin Roche was 15, when he worked for Mnemonics, stands as a good example. One of the technicians Mnemonics hired had been teaching documentary filmmaking at Trinity University, a private liberal arts university in San Antonio. He simply decided that he wanted to go into electronics, and Mnemonics hired him as a technician. Austin Roche had to teach him how to do wire-wrapping on prototype boards.

Phil Ray happened to ask the newcomer if he had access to the documentary film library at Trinity University, and found that he did. So, every day, for the rest of the summer, they spent a 90-minute lunch watching one documentary film after another, borrowed from the Trinity library, until they exhausted the inventory. They were mostly corporate subjects, and Austin Roche's favorite was the one from Alcoa.

Also, they started the blinky-badge contest, to see who could make the coolest badge out of red and green LEDs, using a set number of circuits. They voted to change the rules when they found that the original component list was not enough to let them scroll their names across the LEDs.

He remembered spending all day for weeks on end at the office, sleeping there and returning home only to shower and eat.

"Sometimes I would beg to go home and take a day off, but the boss would say, 'Come back, work harder, we've got all these boards to debug, we've got to do it now.' But we played as much as anything else," Austin Roche recalled.

Indeed, late one night he and his boss decided to see if they could jet-propel themselves down the hall on a roller chair using CO2 fire extinguishers, and collected four from around the office building. Removing the diffuser bolt, they were able to get the exhaust to erupt straight out the back, but on the first try it made the operator's hands so cold he let go. The extinguisher spun through the air, ripping down some ceiling tiles and a neon light, tearing his shirt, and leaving him covered with that appeared to be snow.

They were able to pick the lock of a paper towel dispenser in the washroom to get paper to wrap around the extinguisher. Thus insulated, they were able to propel themselves down the hall and into the wall with the next three.

"We got in a lot of trouble with the building manager," he recalled, unsurprisingly.

On other days they worked to pipe Austin radio stations (considered superior to the San Antonio stations) into their stereo systems.

They discovered that the latest generation of lasers could be turned into powerful hand-held pointing devices, capable of fingering distant cars. Sources disagree about who got arrested and for what.

Jonathan Schmidt stayed at Datapoint, where he welcomed youths who showed up to tinker in the lab. Sitting on the bench stools, sometimes their feet did not reach the ground. He would screen them by showing them an electrical circuit that was mundane, and one that broke new ground, and gauge if they reacted appropriately. He once actually promised a youngster that he could take him to a water park that weekend if he would finish a piece of software by Friday. The youngster did, and Schmidt made good on the promise.

Roche the Person

Gus Roche himself was not a nerd. By all accounts, he acted with the assurance of a man who knew what he was talking about—even when he was making it up.

One time he had to have his Porsche worked on at the local dealership. They were not through by the end of Friday and proposed to leave it on the hydraulic lift until Monday. He told them they should not leave the car on the lift over the weekend, since the hydraulic bushings would fail in the suspension. They looked at him like he was nuts, and ended up leaving the car on the lift. Monday, they took it off the lift—and the car sank all the way to the ground, the hydraulic suspension having failed. The dealership had to fix it, Chris Roche recalled.

One time Roche was in Paris with his wife visiting with the head of Datapoint's European operations, recalled Chris Roche. They went to the manager's favorite restaurant, a small and intimate family-run place in an out-of-the-way location. When they got there they were the only American patrons, the rest being locals. There was an old musket on the wall over the fireplace. Roche—who had studied gunsmithing in his youth—took it down and began examining it. The staff was alarmed but the French Datapoint manager assured them they need not worry—this person was a Texan and knew how to shoot. The staff was enormously impressed and soon restaurant operations were suspended as they went into the basement with him to shoot bottles.

Returning from another overseas trip, his plane landed in Houston, drifted off the runway, and got stuck in the mud. Annoyed by the way the stewardesses were making light of the delay, he threatened to get off the plane using an emergency slide. That motivated them to send for a bus to remove the passengers.

Another time when CBS Evening News anchorman Walter Cronkite and correspondent Eric Sevareid were in town, Roche spotted them at a restaurant. He walked up to them, started talking with them, and ended up spending the evening with them.

He proved adverse to planning for vacations. One time he decided the family would fly to Denver and then rent a car, drive to Aspen, and find a place to stay. His son Chris Roche managed to find a place to stay on short notice in Aspen, but when they arrived at the Denver airport there were no more rental cars available and a storm was moving in.

In a magazine article, the Gus Roche had read the name and hometown of the CEO of a rental car company. He went to the airport counter of that rental car firm and asked the clerk if he knew who the CEO was. He apparently didn't, and Roche supplied the name and a phone number.

"Call him and tell him that you don't have a car for Austin Roche," he said (using the name on his ID rather than his nickname.) Instead, the clerk immediately found a car. As for the phone number, it was a bluff—the area code was correct for the hometown, but the rest was made up.

He rose to the occasion when his family life got ensnared in the dark side of computing—or what passed for it at the time. Two of the Roche children took advanced mathematics classes at Alamo Heights High School, whose nationally recognized teacher had convinced the school to acquire a subscription to a time-sharing service. So the school ended up with a Teletype machine that accessed a data center in Kansas City via a telephone modem. The students would use it for Basic programming, remembered Chris Roche. The school was also using it for various back-office functions. At the Roche home, there was a Datapoint terminal in the family room, with a modem and a WATS[95] line. On a rare San Antonio weekend when it was raining, the younger generation had no trouble using those resources to access the data center in Kansas City that the school used. They got into the school's files, left unprintable comments about various teachers, and tried to change grades.

Word soon got out that the school officials had found the hacking and were unhappy. One of the boys who had taken part—who later became a physician—began singing like a canary, and it all came down on the head of Chris Roche, as the telephone connection came from his house.

"I was called into the office," he said. "I told about how we have an open house, and how kids hung out there all the time as mother would take them in when they had to get away from their conservative parents. I dodged it all."

---

95 Now obsolete, WATS (Wide Area Telephone Service) lines were dedicated long-distance lines sold at a discount.

Gus Roche subsequently smoothed things over by giving the school a Datapoint terminal on free loan, so they could upgrade from the Teletype.

As for the time-sharing service in Kansas City, the concept of intrusion and hacking was evidently new to them. "They were amazed," Chris Roche said. "They gave the school free time because they were aghast that kids were able to find their machine, sign onto their service, and start messing stuff up. They were pleased as Punch that it was Alamo Heights High School that suffered the damage, and not some large corporation."

Engineering the Engineers

But most of the memories Gus Roche left behind involved socializing at restaurants.

"Five nights a week my father would be holding court at restaurants," Chris Roche recalled. "After the place would close the owner would let him keep going. All sorts of people could be there, including visitors from out of town, people associated with the business, and other engineers. Phil Ray would sometimes be there. Sometimes I would put on a jacket and hang out there, to spend some time with my dad and feel older than I was. The discussions would be wide-ranging, and not just about business. He was generally the focus, and the authority."

But there was more going on under the surface—despite appearances, he was not there to socialize. He was there to implant his vision on those around him, or to establish what their vision was, and, if it seemed suitable, nurture it. He did this relentlessly, and, frankly, most memories of him were evidently generated by some expression of that effort.

"Simply socializing was not the method he used to get information," explained Austin Roche. "It was much more invasive than just getting them drunk and asking questions. If he sensed that you had hidden agendas, or you were not performing as well as you could, he had to work on you. If he could not read you immediately he had to work on you. If he read you well and saw you were being hindered by someone else, like an engineering manager who would not allow you to pursue your vision, he had to work on you.

"He was a big believer in, 'Let's go out and eat.' It was a big statement that there was something significant going on here, that this person was important right now. He loved the mystery that it created."

At the table, he could exude the charisma and start pushing, Austin Roche explained. "If they were not doing enough he would say, 'I know you can. What are your ideas about doing this? Aren't you thinking this? I'm sure you'd be thinking this. If you were really in the groove you'd be thinking what's next, and this is what's next.'

"Or, if they were doing good, he'd say, 'You could do more. There is so much more you could do. You can be the head of manufacturing. You can be the head of engineering. What's holding you back? What's your problem? What's in your way?'"

Flattered by the attention, the subject would usually open up about his thoughts and feelings.

"Then he'd say, 'Aha, now we're making progress. Now we're getting someplace with you.

Now we're starting to talk. This is what I love about you. This is good,' he would say. Then he would seize the moment. 'You need to do this all differently,' he would say. 'You're not thinking about this right. I could tell, and I was just waiting for you to tell me. You need to do this instead.'

"Frequently, what he told them next was controversial," Austin Roche explained. "He would say things like, 'I think you're better than the engineering manager. So what are you going to do about that? I'm not going to tell you how to do it, but you need to go after that guy's job.'"

He would also call a person in and ask about their vision, and tell them if he thought it was right or wrong. "If you would not share your vision, he would discount you, and say he was disappointed in you. You had to open up if you wanted to be close to him, and on the inside track. Otherwise, he'd assume either that you didn't have anything inside you, or that you had no more to show. That did not make you a bad person or limit your role in the organization, but he'd say, 'I'm looking for other kinds of people.'

"He did it to me. He did it to everybody," Austin Roche remembered. "You were expected to excel and achieve beyond your wildest imagination, but not based on any formalized training, rules, or methods. The more unconventional the achievement was the more he respected it. Irreverent, unconventional, and dynamic is what interested my father, because that is how he was."

His approach brought out three kinds of reactions: those who were motivated as intended, those who were bemused, and those who did not understand.

"The ones who were motivated rose to greater heights than they could have imagined," Austin Roche explained. As for the bemused, they were more secure in themselves and would simply play along. But those in the third group did not always do well.

"Some fell apart because they really could not take the scrutiny," he said. "They got really scared and did not understand what he was doing at all. Dad would say he felt really terrible. 'I destroyed this guy, but I was not trying to destroy him, I was only trying to motivate him.' He would offer them different career options.

"'I can't take it! I can't take it!' I remember one guy yelling at my dad. 'I've had all I can take, I've got to go, I'm leaving.' And he walked out the door and never came back. My dad said, 'Oh God, I was just too rough on him. I had too high expectations for him. I just didn't know.'"

Part of his approach could be traced back to his experience as an engineering manager in the space program. There, Roche had to rapidly bring entirely new technologies to life. The process called for the very best that each individual had to contribute. Bringing out the best in a person involved that person recognizing, shaping, and applying his or her personal vision.

But in hindsight it is clear that there was another element behind Roche's intensity. It was about to come to light.

Requiem

Jack Frassanito had dinner with Gus Roche on Monday, February 10, 1975. It was not one of Roche's signature vision sessions, as Frassanito found him distraught about the overall situation. Being pushed out of Datapoint had been disheartening, and now he found himself all over again in a new startup with Phil Ray. Roche feared this was a mistake, Frassanito later recalled. Roche also mentioned being sick, and dreading surgery.[96]

The next night, Tuesday, February 11, Roche went to dinner again, at a steak and lobster restaurant[97] beyond what was then the north edge of town on San Pedro Avenue. This time he was there to listen to someone else who was disheartened, iconoclastic Datapoint manufacturing executive Dick Norman, who had been with the firm almost since its founding as CTC. Norman had a run-in earlier that day with O'Kelley. Roche did not do any drinking.

Roche headed home by himself after 1 am, meaning that it had become the morning of Wednesday, February 12, 1975. He was driving a new blue four-door BMW—not the blue Porsche whose acquisition caused a furor five years earlier.

After driving about five miles south on San Pedro, he turned east on Sandau Road, which runs parallel to the southern perimeter of the airport. Had he turned west, he would have been on Rhapsody, and in a few blocks would have passed the original CTC headquarters building. The area was not new to him, in other words.

After about three-quarters of a mile east of San Pedro Avenue, Sandau Road crosses Isom Road and abruptly turns 45 degrees to the right, officially becoming Jones-Maltsberger Road in the process.

Gus Roche missed that turn. He apparently swerved to avoid a utility pole on the other side of the intersection but instead slid sideways into it, impacting on the driver's side. He was wearing a seat belt and later tests showed that it was functioning. He nevertheless twisted to the left and suffered massive injuries on the right side of his head.[98] There were no witnesses.

He was taken to a hospital in critical condition. He never regained consciousness and died of his injuries on Saturday, February 15.

In the meantime, those members of the family who had not already been carrying the burden of the secret were told—he had had terminal stomach cancer. He had not been expected to live for as long as he had.

In other words, his on-going efforts to probe the people around him and motivate them by kindling some personal vision within them had been done, at least in the last couple of years, in the full knowledge of his own impending death.

Funeral services were held Tuesday, February 18. The police escort was alarmed by the

---

96 Interview with Frassanito in 2008.
97 Casey's John Charles, which no longer exists.
98 Details are from interviews with Chris Roche and Jack Frassanito, and from the Express-News archives for February 13, 1975. Police archives for 1975 were not retained by the department.

length of the funeral procession.

At Mnemonics, the original money had run out, and the investors were not willing to put up more with Roche gone. Ray also lost interest. The staff was able to make some working chips—the first CCD memory devices—but the company was dissolved by the end of the year.

"The whole period was intense and father was the focal point, Things ran at 90 miles per hour based on his vision and his faith—and bang, it was gone," said Chris Roche. "People were left stunned and agape—they never thought it would end."

Chapter 12

The Microprocessor Market Blossoms: 1972-1977

While CTC was becoming Datapoint and deciding it did not need Intel processor chips, Intel was doing just fine. The chip it made for CTC/Datapoint launched it onto an entirely new path, with results beyond anyone's imagination.[99]

The 8008 chip was released to the market on April 15, 1972, with a retail price of $120. This event was probably not noticed at CTC, which had recently again avoided bankruptcy, and was installing the first Datapoint 2200 desktop computers.

Hobbled by the use of 18 pins, the 8008 was not really a computer-on-a-chip, since it took three dozen support chips to multiplex the input and output signals of the data pins that they shared. But, once those chips were in place, plus some memory circuits, some kind of keyboard, and a display, and a storage system, it was indeed a computer. And that, oddly enough, was a problem for Intel's upper management, since they were terrified that the computer makers who bought Intel's memory chips would start seeing Intel as a competitor.

Even a year later, in 1973, sources recall Gordon Moore becoming incensed when he found that the marketing staff planned to promote the chip as a computer at an industry conference. He ordered them to do no such thing, since it would put Intel in competition with its primary customers. They displayed the chip anyway, but removed all blatant computer references.

Robert Noyce, on the other hand, became an early convert, and pushed for the chip to succeed. He predicted (accurately) that microprocessor chips would one day be as common as fractional horsepower electric motors (such as are found in appliances and power tools.)

But, as it turned out, the customers that initially adopted the chip were not making computers but stand-alone intelligent devices. In hindsight, that made sense—the computer makers at the time already had their own custom processors and (like Datapoint itself) saw no reason to turn to outsiders. The device makers, on the other hand, wanted to add maximum intelligence with minimum fuss, and a tiny, inexpensive computer let them do so.

Consequently, the first uses were in scientific calculators, blood analyzers, machine tool controllers, traffic light controllers, point-of-sale devices (i.e., electronic cash registers), and check processing machines. Not only did this development spare Intel from the wrath of its primary customers, but it turned out that the controller market was then easily ten times larger than the computer market. Interest was so high that for a while Intel made more money selling reference manuals for the chip than from selling the chip itself. (They were the first in the industry to charge for technical literature, incidentally.)

Hal Feeney switched from engineering to marketing in early 1972 as Intel prepared for the 8008 launch. "When we first announced both the 4004 and the 8008, they were presented

---
99 Information in this chapter is derived from author interviews with the quoted individuals, with background information from transcripts of interviews conducted by the Computer History Museum in San Jose, California.

as a new concept," he said. "At the time system designers were really logic designers who put logic blocks together on individual circuit boards. One of our roles was to educate them on how they could program their logic using a microprocessor. We spent a year and a half selling that concept. It was amazing—it really struck them, and changed the way they operated entirely. They had been solving specialized problems where they often had to make their own complex designs.

"These devices were really too slow to compete with the computers being made by companies like DEC. Intel got into the market in stealth mode. As performance increased the chips became more competitive, but that took a number of years.

"You could say that Intel was lucky to get into the microprocessor business, or that it made its own luck. After Busicom went under and CTC turned out to be not interested, Intel would have had two designs that were little more than boat anchors had they not been designed for multiple applications," Feeney said. The 4004 series was the forerunner of the embedded system industry, while the 8008 was the forerunner of modern microprocessor-based computer, he added.

The adoption of the microprocessor chip by device-makers was facilitated by the parallel development of EPROM (erasable programmable read-only memory) technology. For a stand-alone device, the software had to be permanently fixed in ROM, rather than, say, loaded from a disk. With an EPROM, the programmer could debug and create the ROM for the prototype at his or her desk. Then, if production numbers were relatively low, the programmer could immediately make copies for the production models. The use of EPROMs could save three months, Feeney recalled.

Before the 8008 even came out, in late 1971, Intel sent Federico Faggin to Europe (presumably because he was European) to ask key customers what they thought about both the 4004 and the 8008.

As for the 8008, those who were looking for a solution appreciated it, he recalled. Those who were in the computer business tended to express hostility to the idea, poking holes at its functions, and complaining that Intel was getting into their turf.

"The English were the worst," Faggin said.

But there was also constructive criticism from various quarters. Beyond objecting to the need for dozens of support chips, they thought the speed was too slow and the interrupt handling was weak.

Faggin proposed a revised version. Intel's response was to assign him to develop a coin recognizer for a snack vending machine.

Six or seven months went by. During that time the 8008 came out and Intel decided there really was a market for it, so Faggin got the green light for the follow-on chip. It required 6,000 transistors, and they called it the 8080.

Faggin set out to use the new N-channel fabrication technology, which produced circuits that were about twice as fast as those produced by the P-channel technology used in the 8008. Faggin brought over Masatoshi Shima to help. By this time Shima was working for

Ricoh and the head of Intel had to get formal permission from the head of Ricoh to hire him.

Starting work in mid-1972, they had the first working chip at the end of 1973. It was software compatible with the 8008, although Faggin added some instructions. A recent custom chip project at Intel had involved a package with 40 pins, so Faggin was also able to use 40 pins, ending the shortage of output pins that bedeviled the 8008. Consequently the 8080 only needed about six support chips. Altogether it was about ten times faster than the 8008. It had 16-bit addressing and could use 64K of memory, instead of 16K of memory as in the 8008.

It came out in April 1974, two years after the 8008, and was priced at $360.

Faggin states that the 8080's design was entirely his own, or based on customer feedback. The Datapoint engineers do not recall it that way.

According to Jonathan Schmidt, during the development of the Datapoint 2200 version II (1970) Datapoint started feeding Intel additional instructions that we really needed to make the processor more useful, such as to increment the H and L pointer by one. In the end there was only about a 2 percent difference between the 8080 and the processor architecture of the Datapoint 2200 Version II designed by Harry Pyle and Vic Poor, although that was enough to make their software incompatible. Datapoint's developers came up with some add-on logic that would bridge the gap and make the 2200 Version II compatible with the 8080, but it was never used in a commercial product. By the time the Intel 8080 came out (1974), the board level 2200 Version II was long in production and Datapoint was developing its next-generation (5500) processor design. Consequently, the appearance of the 8080 mostly confirmed our belief that Datapoint was at least 18 months ahead of the Intel 8080 architecture Schmidt indicated.

Intel, however, soon found that it was barely ahead of its own chip-based competition. Six months after the 8080, Motorola came out with the 6800 microprocessor chip, ending Intel's monopoly. (Faggin noted that Intel would have had an additional six months if his proposal for the 8080 had been approved immediately.) It too had 40 pins and a 64K address space, but had an entirely different instruction set.

It became a three-way market in September 1975 when MOS Technology came out with the 6502 microprocessor chip, which was similar to the 6800. (It had, in fact, been designed by some of the same people.) Other processor chips, now mostly forgotten, followed in a steady stream.

Nevertheless, sales of the 8080 were so strong that Intel recovered the chip's development costs in five months. (The appearance of the 8080, incidentally, did not end demand for the 8008. Vendors who had designed it into their products could not switch to the 8080 without a major redesign, if only because it had different pin-outs and support chips.)

$2,000 per Day

While the 8080 was under development Faggin was assigned several other development projects, including one related to one of the first LCD digital watches, which was a financial bloodbath for Intel. Eventually he had 20 projects and a staff of 80 people—and the previous startup ambiance was lost amidst layers of bureaucracy.

In early 1974, Intel began monitoring when workers showed up in the morning, and everyone who got there after 8 a.m. was reprimanded. This alienated Faggin, who got no consideration for the fact that he worked nights. Then he found that Intel had patented one of his ideas without telling him. He decided he would be better off setting up his own firm, and left Intel on Halloween, 1974.

He formed a partnership with another former Intel manager, Ralph Ungermann, and sent a press release to Electronics News saying they planned to form their own microprocessor company. This led to him being approached by Exxon Enterprises, the investment arm of oil giant Exxon Corp. They liked his plan to produce a processor chip that embodied certain improvements over the 8080.

Funding began in April 1975. They called the startup Zilog. Faggin was able to hire Shima again, and they had working samples of their new chip, called the Z80, in February 1976. They were able to begin selling the chip in July, initially using third parties to fabricate it.

In three more years the company had 1,000 employees and Faggin had surgery for a perforated ulcer.

In 1981 he had Exxon buy him out, based on the valuation that Zilog[100] had by then achieved. He put the money in the bank, where the interest amounted to $2,000 per day.

He would go on to other ventures, but his work with the early microprocessors would have extra meaning for him—and would be a perpetual source of annoyance for Intel.

"Intel likes to rewrite history," he complained. "They could not possibly admit that the 8008 architecture was not theirs. History, as far as Intel is concerned, begins with the 8080. The 8080 gets more air time then the 8008 by a factor of 100 to one," he said.

Looking back at the genesis of the 8008 and the 8080, "Those were very special times, when the technology came together and you could do things like this," Faggin said. "If you had thought of it ten years earlier, so what—you couldn't do it. But we were able to catch the wave at the right moment, and if we hadn't someone else would have—it was inevitable. Silicon-gate technology positioned Intel to do it a couple of years before the rest of the pack, and a couple of years later it could have been done with metal-gate technology."

The current digital world is derived from what was created at that time. But Faggin added that the results were not then foreseen. "I would not have believed what we have today if you had showed it to me," he said. "A computer on a desktop made sense for some applications, but if you had told me that everyone and his brother would have a computer even at home, I would have said you were full of shit. If you had said you could Google something on the go, I would have said you were really full of shit."

---

100 At this writing Zilog Inc. was still in business, although under different ownership. It still made the Z80.

The x86 Dynasty

Back at Intel, the 8080 was followed in 1977 by the 8085. It used less power and in certain settings required no support chips, making it the first true single-chip processor. It was software compatible with the 8080. It became more popular in self-contained microcontrollers than in desktop computers.

More significant was the next member of the dynasty, the 8086, released in 1978. It used a 16-bit architecture, with memory segmentation hardware that permitted it to address up to a full megabyte of memory, albeit in 64K segments.

The resulting system was referred to as the x86 architecture, and became the foundation of the PC software industry. However, its instruction set was expanded, and was no longer directly compatible with the 8008 and 8080. But software for the older chips could be re-assembled into code that could run on the 8086, since its instruction set included equivalents of the previous instruction sets.

Then came the 80186 and the 80286 in 1982, the 80386 in 1985, the 80486 in 1989 and (finally dispensing with numeric names) the Pentium in 1993. Each was faster, more powerful and feature-rich than the previous one. (Each was also cloned in bulk by other chipmakers.) With the first Pentium the transistor count was up to 3.1 million. The Pentium line was regularly enhanced for more than a decade, peaking with a version of the Pentium 4 that came out in 2006, which had a clock speed of 3.8 gigahertz (i.e., 3.8 billion cycles per second) and used 118 million transistors. It could directly address 64 gigabytes of memory—more RAM than some operating systems knew what to do with.

Since then, emphasis has shifted to multi-core processors, which put the equivalent of more than one computer processor on a chip. The switch to multiple cores resulted from the discovery that clock speeds higher than about 4 gigahertz caused processors to run afoul of heat dissipation problems. (In other words, they glowed red and then failed.) Henceforth, performance increases will have to come from clever programming using parallel processing, rather than simply relying on the hardware to get faster and faster. At this writing the jury is still out concerning how practical that's going to be.

But for the purposes of this history, the most important Intel chip was a variant of the 8086 that came out in 1979. Called the 8088, it used an 8-bit external data path instead of the 16-bit path used by the 8086. (Internally, it was still a 16-bit machine.) Its reliance on 8-bit I/O reduced throughput, but let it work with cheaper 8-bit circuit boards.

As will be shown, this otherwise uninteresting device spawned an industry that eventually crushed Datapoint.

Chapter 13

Invention of the LAN: 1975-1977 (and beyond)

As the 21st Century rolled through its first decade, an office without PCs was inconceivable, and it was inconceivable that those PCs would not be networked. In a large office they would share their files and exchange e-mail within and outside the organization through a server. The technology that allowed such activity is called a local area network, or LAN. Even small offices have LANs, if only to connect the computers to a high-speed Internet connection.

Like the microprocessor, the LAN was also invented at Datapoint. Unlike the microprocessor, there was no intermediary like Intel or Texas Instruments involved—the idea originated at Datapoint, was brought to fruition at Datapoint, and was successfully marketed by Datapoint. With the LAN, Datapoint's role as a pioneer is undisputed.

But as with the microprocessor, the fruits of that invention slipped through Datapoint's fingers.

Office automation consultant Amy Wohl of Pennsylvania remembered 1987 as being the crossover year for how office networks were perceived. "At that point so many office workers had computers as workstations that e-mail went from being an interesting idea to something people were implementing all over the place. Before that, as soon as the client found that they had to have a computer for every office worker in order to have e-mail, the numbers got scary for them. But as PCs became ubiquitous all you had to do was hang a LAN, and along with e-mail came other white collar applications like calendaring."

But for Datapoint office automation users, LANs and office e-mail was a part of their environment ten years earlier, as Datapoint unveiled the first LAN for desktop computers in 1977. Using Wohl's standard, Datapoint was ten years ahead of its time.

Establishing the Need

Computer firms at the time had to innovate constantly to stay competitive, but just how they should apply their product development resources was always an open question. Most were eternally seeking a balance between being market-driven and being research-driven. Market-driven meant that they were coming out with new products based on the stated needs of their customers. Research-driven meant they introduced products that their engineers thought would be interesting. Stock analysts typically dismissed the research-driven approach, equating it to self-indulgence by the engineers. Corporate management would pay lip service to that opinion—but also typically realized that a market-driven approach would just lead them into doing the same things over and over again. For instance, if Bill Gates' efforts had been market-driven, he would have spent his career producing more and more refined versions of his first product, Microsoft Basic.

In the case of the LAN, the customers were not demanding any such thing in 1977. They couldn't have—they'd never heard of a LAN. Sharing files meant passing around removable media, like tapes. Consequently, the first LAN was a premiere example of research-driven product development.

"The idea came from the development group that I was running," said Vic Poor. "I thought it was far-fetched. I had other priorities at the time and did not want to fool with it. But fortunately the people who worked for me paid no attention and went ahead and did it. It was altogether an engineering-driven idea, the customers had no clue, and no one was asking for it. That was why it was not given a high priority."

In fact, the idea came from Harry Pyle, who could recall the exact moment when the idea came to him—he was with a small group of Datapoint colleagues at an Italian restaurant, having a meatball sandwich for lunch. "One of the field engineers was saying that this DATASHARE was great, but we need to have a lot more terminals on it, and the customers wanted it now. Even at that time, other engineering groups were trying to figure out how to make bigger and bigger 5500s, but they needed the product now."

DATASHARE, as explained in Chapter 10, had become the mainstay of Datapoint's product line. A processor (originally the 2200, and by now the 5500) would be connected to multiple "unintelligent" terminals such as the 3300 and its variants. Sitting at each terminal would be a person, typically a clerk, who was doing data input (such as payroll or ledger updating) or data lookup (such as inventory checking) or some combination.

The user sitting at each terminal had the illusion that he or she had full control of the computer, because the processor was fast enough to switch its attention between each of the users and respond to their keystrokes without off-putting delays. Or, it could if there were not too many terminals demanding attention at the same time.

Just how many terminals could be supported at one time depended on the complexity of the software they were running, but Pyle recalled that, on a 2200, they usually hit the wall with four or five simultaneous users. If more terminals could use the same machine, more clerks could access the same databases. When there were too many clerks for the machine to support, the only way to keep the customer happy was to offer a bigger, faster machine that would support more terminals. But if the customer was already using the biggest machine Datapoint offered, the only alternative was to try to make do until Datapoint produced an upgrade—unless the customer switched to another vendor who offered a bigger machine.

Having as many clerks as possible being able to use the same machine simultaneously was important because, in applications like inventory management, input from one clerk (such as adding an item to the inventory) would immediately affect what the other clerks could do. (They could, for instance, then sell the newly added item.)

On the other hand, if one machine could access the disk files on another machine in real time, then the clerks using the terminals on the second machine might as well be sitting at terminals attached to the first machine. Instead of throwing out the inadequate machine and porting everything to its new, bigger replacement, you could just add a second machine, and then a third, etc. Capacity could be smoothly adjusted to meet demand.

But first, they had to get those machines to talk to each other in a manner that was fast and reliable.

"We were using Winchesters[101] with removable disks, and the data rate off the disk was about 2.5 megabits," Pyle recalled. "If we could communicate between multiple machines at speeds comparable to the speed of the data coming off the disk it would be like one disk connected to multiple machines. So I started thinking about ways to send the data between machines. I had read an article about the IBM 360 I/O channels, with data over fat cables the size of water pipes. That inspired me to think about serialization schemes (i.e., to avoid fat cables) and 2.5 megabits was considered pretty fast."

They went ahead and settled on a speed of 2.5 million bits per second. To carry that signal they used coaxial cables similar to those now used for cable TV hookups, but more importantly matched those that were then used to connect IBM 3270 mainframe terminals, which were at the time the most commonly used computer terminals in the industry. They decided to adopt the same cabling rules as IBM, which limited cable runs to no more than 2,000 feet.

Token Passing or Collision Detection

Getting the machines to talk to each other reliably was a thornier problem. Left to themselves, each computer would have things to transmit at random moments, and inevitably two or more would want to transmit at the same time. The simple transmission system they were using did not allow multiple frequencies, so simultaneous transmission by multiple machines would result in useless line noise. There had to be some way for the machines to take turns, and do it automatically.

The first method that suggested itself was called collision detection. It can be compared to two people arriving at a narrow doorway from the same side. Seeing that they both can't pass through the door at the same time, they throw dice to see how many paces backward they should take. They then resume walking toward the door and the one who has stepped backward the least naturally goes through the door first, unimpeded, followed by the other.

Following that analogy, a computer can start transmitting if it notices that the line is quiet. If it notices that another computer started transmitting at the same time, it will stop and wait for a random number of microseconds. After this fallback, it will try again—unless the other machine (having randomly waited a shorter interval) has resumed transmitting in the meantime. Then it will wait for silence and try again.

(To keep one machine from hogging the wire indefinitely with a long transmission, data that is to be transmitted is chopped into small data packets. Each machine would be jockeying to send a packet, rather than an entire file.)

The collision-detection concept seems straightforward in theory, and when using the two-person analogy it even seems foolproof. But instead of two people, what if a crowd is heading toward the door? At some point most attempts to get through the door will result in collisions, and much more time will be spent rolling dice and falling back than in the act of moving through the door. In fact, there would be no way to predict in advance just how long it would take any specific person to get through the door.

---

101   Rather than referring to a rifle, Winchester was used as a generic name for a kind of hard drive.

What killed it for the Datapoint designers was the idea of trying to troubleshoot the system if something went wrong, when the system was built around random factors (i.e., the fallback intervals.) Pyle remembered the designer, John Murphy, pounding on a table, saying he would not build randomness into his product.

Instead they came up with a polling scheme, which was slower, but promised fully predictable results (i.e., it was "deterministic.") An analogy would be a group of hyper-polite people standing in a circle, each waiting for a turn to speak. A person can speak only after being handed a baton by the person to his or her immediate left after that person has had a turn to speak (and has additionally received an acknowledgement from the person spoken to.) So the baton is continually handed around the circle. When getting the baton, a person can pass it to the next person in silence if there is nothing to say. Or the person can say something, wait for an acknowledgement, and then pass the baton.

In the actual computer system, the baton was small message called the token, and the scheme as a whole was called token-passing. Instead of being passed around a physical circle, each machine was assigned a random eight-bit ID number (initially with hardware switches) and each would pass the token to the machine with the next highest number. When a new machine joined a network, it would send out a burst of line noise to "kill the token." The machines would then go through a reconfiguration routine and discover the number of the next computer in line. They would use a subset of this procedure, without having to kill the token, when they noticed that a machine had dropped out.

Poor recalled that the basic idea came from an old textbook on telegraphy, with its description of how an operator of a Teletype machine on a network with other Teletype machines could use a "break key" to generate line noise and force the other machines off-line.[102] After silence reigned, the operator who had used the break key would then start transmitting.

Hubs and RIMs

Using eight-bit binary ID numbers, there could be 255[103] machines sharing a network. Since the designers could not just run a coax from every machine to every other machine, the designers had to agree on a "topology." The arrangement they ended up with is usually called a "star topology."

Each computer connected to the network through an attached box that handled the networking chores using a rudimentary processor called a sequencer. System designer John Murphy[104] recalled that the circuit ended up including a random number generator in a non-standard fashion simply because one had been left on the prototype's breadboard after of a previous design experiment. It has remained in the design ever since, he indicated. Likewise, they used 256-byte data packets for transmissions, to match the 256-byte disk sectors then in use. However, they later realized that some bytes in each data packet had to be used for an address label, so that a full-length disk sector overflowed the data packet and had to

---

102 Its descendent is the PAUSE/BREAK key on the top right of the standard PC keyboard layout in the U.S.
103 With an eight-bit ID number you have 256 combinations, since two raised to the eighth power is 256, but the zero value was reserved for broadcasts to all machines.
104 John (Murf) Murphy was interviewed in 2011.

be sent in two packets. They could have made the data packet larger but that would have required more memory and more expense, which at the time was not practical, he noted.

Originally this attached box was called a BAIL (Bit Assembly Intercom Link) and then PAIL (Pass Around Intercom Link) and then a FRIL (Fast Resource Intercommunication Link.) Naturally, with the last one there were those who said, "Just add a few frills and you've got a network." Perhaps wisely, the name was later changed to RIM, for Resource Interface Module. Originally each RIM was about twice the size of a breadbox and usually sat on the floor under the desk that carried the computer. Since a RIM was connected to its computer through an existing data port, existing Datapoint computers could be networked without hardware modifications.

If there were only two computers on the network (as in two computers running DATASHARE, each with multiple terminals, sharing a database on one of the computers) they could just ran a coax cable between the ports of the RIMs of the two machines. This was called a point-to-point configuration. But the idea was usually to have more than two machines, and the way to connect more than two machines was with hubs.

A hub was simply a box with multiple coax ports. A signal arriving in one port was repeated out the others, so every computer connected to a hub heard the traffic of all the other computers. Hubs usually had four, eight, or 16 ports. If a network had more than four, eight, or 16 computers, one port on a hub could be connected to a port on another hub, expanding the possible network connections until the maximum of 255 computers was reached.

The four-port hub was also called a passive hub since it was just a connection box. The larger hubs were called active hubs and regenerated the signal, but they required an AC connection.

The official maximum cable length was, as mentioned, 2,000 feet, but by using ten active hubs a connection of 22,000 feet could be sustained between two RIMs. With later modifications a network could cover 74 miles end-to-end. These far exceed the range of any other LAN technology.

Software Support

Simply being able to communicate with other computers was not enough. The idea was for one computer to use the disk drive of a second computer, as if that drive on the second computer was actually built into the first one, instead of being located across the room or the building. Moreover, there had to be a way for that shared drive to be used by more than one person at a time. If everyone who wanted to access a file also wanted to make changes, the result could be chaos.

Schmidt credits Gordon Peterson with this feat. Peterson started working for Datapoint in 1974, but had been working with Datapoint hardware for the previous year while he was as student at the University of Illinois at Champaign. A professor there reached across his desk and handed Peterson a 2200-programming manual. Intrigued, Peterson was subsequently able to borrow a Datapoint 2200 to work with.

"I thought it was a really neat, cool machine," Peterson recalled. "It was the first personal computer worthy of the name, where the computer was really dedicated to the person sitting in front of it. Even with the PDP-8 there were typically teams of people using it, so to have a machine dedicated to your use was pretty special.

"It was a fully functional data processing system, suitable for accounting and all types of stuff. DATASHARE let you do large decimal math that was precise to the penny. A lot of minicomputers used floating-point math, which led to round-off errors so that you would lose pennies here and there. Datapoint also had unusually good communications support for the era, in that you could send and receive files using high-level languages," Peterson said.

He met Schmidt at a computer industry convention in April 1974, and showed him some of the programs he had written. A day or two later he flew to San Antonio and picked out an apartment, signed a lease, and returned to Illinois to pack. He started work on May 1, 1974, he recalled.

Peterson explained that, to avoid chaos, a multi-user computer system needs a procedure called file (or record) locking. When a user reads a file for the purposes of modifying it, the system needs to reserve it for the exclusive use of that person. When that person is through, the system then needs to release the file. (No locking is needed for someone who wants to read a file with no intent to modify it.)

DATASHARE, being a multi-user system, already had a file locking procedure, but when expanding it to a network the trick was to avoid a situation commonly called "the deadly embrace" but which Peterson preferred to call the "indefinite postponement problem."

"It's where two users have the same set of resources," he explained. "The first user gets resource A and is waiting for resource B, and the second has B and it waiting for A. The way you solve that problem is that you require both users to state the complete set of resources they need, and system sets them up in a canonical (well-defined) order, to make sure that you always allocate things in the same sequence."

The upshot is that the necessary functions were present in the Datapoint operating system by the time its new LAN was finalized. When users upgraded to a LAN no reprogramming was required for their applications. All they had to do was add the RIMs, hubs, and cables. Disk files anywhere on the network could immediately be used by anyone else on the network. Files could be accessed by their address on a given disk, or the system could go out and find it by name anywhere on the network—a feature that, more than three decades later, Peterson was still waiting to show up on a Microsoft operating system.

Beyond that, "Before we were finished developing ARC we thought, 'It can't be this easy. What are we overlooking?'" Peterson recalled. "We thought, 'If it is this easy, why has someone not done this yet?' Every time we went through that, we decided it was that easy, that it will be that neat, and that no one had gotten around to doing it yet."

The Internet?

Naming a new product can be difficult, especially when it represents an entirely new concept, as the effort to name Datapoint's network was no different. Peterson recalls that the name originally used for it was Disp-DOS, for Dispersed DOS. They then switched to DataNet, but discovered the Honeywell as already using that name for a product. Then they switched to DOSNet. This didn't last either.

Amazingly, a Datapoint interoffice memo dated November 1, 1977, concerning plans for the official product announcement a month away, referred to the product as the Internet. This could not have been the official name for long, and some sources were unaware that it was ever used. The thing that is now called the Internet was then a collection of government and academic computers called the Arpanet. What Datapoint envisioned was a purely local network, confined to one building or campus, rather than a worldwide network, but both its network and the modern Internet are based on the idea of remote, shared file access.

As for the Internet name, "I vetoed that," said Schmidt. "In the computer business, 'network' meant something slow and unreliable, and I did not want 'net' to be part of the product name."

Instead, he was reading computer magazines on a flight to France when the phrase "attached computer" jumped out at him. "I envisioned an 'attached resource computer' that could let any resources be attached. I wanted to emphasize 'resource,' to keep them thinking about what resources could be attached—there was a whole universe of things that could be useful."

And so it was dubbed ARC, for Attached Resource Computer. Datapoint's press materials for the initial product announcement did not use the word "network," and press coverage followed suit.

ARC was officially announced as a product on December 1, 1977. The first installation was at Chase Manhattan Bank[105] in New York City, begun in September 1977.

"When we took the first ARC to Chase Manhattan, they said they had never seen such as significant product announcement that involved so little hardware," Peterson recalled. (Only RIMs, hubs, and cables were needed.)

Peterson also recalled that the Chase Manhattan installation was for an internal money transfer and letter of credit system. "They had been a DATASHARE customer and had done some amazing stuff, actually, but they wanted more than what a 6600 could do. They wanted to hang more terminals off it and run more jobs. ARC let them stay in the Datapoint product line. Suddenly we could grow with them as their needs grew by adding more processors."

Other customers apparently agreed. After the announcement, "What amazed me was how sales went through the roof—we found tremendous interest in the user base," said Poor. "I had no expectation of that. I thought it would be a hard sell, since it was all new and different. We thought that the security implications of sharing files would be pretty scary.

---

[105] In 2000 it became JPMorgan Chase.

But it turned out that the customers weren't worried about anything—they just took them. We thought the limit of 255 computers per network was out of sight, but a couple of big firms starting pushing it immediately. But each computer could be daisy-chained to more than one network."

Wohl remembered seeing her first ARC at Datapoint headquarters shortly after the 1977 announcement. Previously, getting a high-speed connection between two computers was a complex task that involved getting a dedicated data line from the phone company, and it was rarely done outside large corporations, she explained.

"Doing it inside an office in an easy, relatively inexpensive way was a new idea entirely. We understood that we could attach wires more easily, but I am not sure we understood what that meant in terms of the applications that could run on top of the network. We had to use it and see what it could do," she said.

"Think of how casually we set up networks now," marveled Wohl. "I remember having a network put in at the Atlanta convention center to run a videoconferencing demonstration in 1980. I had to get a grant from the phone company since it took $100,000 to run the lines."

In 1980, after the market acceptance of ARC seemed assured, work began on a single-chip version of the RIM box, so it could be built into a desktop computer. Still smarting from their failure with Intel to get a custom chip made, they set up a team to design it in-house, aiming at a design that any chip foundry with up-to-date processes could make. They were successful with the design, called the Newport chip, but then the challenge was to find a chip foundry that could produce the design reliably. Two foundries tried and delivered yields as low as one die per wafer. One foundry delivered samples that had a short in them, and the next batch was not due for six weeks. The engineers put the samples under a microscope and found that, with microscopic probe wires, they could put a scratch in the right place to fix the grounding, allowing them to at least test the design. Eventually, they approached Standard Microsystems Corporation (SMC) of Hauppauge, NY. SMC agreed to solve the production problems in return for being named Datapoint's exclusive supplier of the chip, and the right to sell the chip. The chips became available to the public through the SMC catalog in 1982.[106]

First, But Not Last

Despite the positive start, there was no happy ending. ARC would remain an integral part of Datapoint's computer product line for the rest of its history, but that did not keep ARC from being pushed aside by another LAN technology.

While ARC was under development, the Datapoint engineers noticed an article[107] in the July 1976 volume of a technical journal called the "Communications of the ACM." It was titled, "Ethernet: distributed packet switching for local computer networks," and described the development of a "broadcast communication system for carrying digital data packets among locally distributed computing stations" at the Xerox Palo Alto Research Center (PARC)

---

106 The Newport chip story is from the Fall 1998 issue of ARCNETworks, the newsletter of the ARCNET Trade Association, and personal communication with Michael Fischer.
107 See http://portal.acm.org/citation.cfm?id=360253&coll=ACM&dl=ACM&CFID=50907435&CFTOKEN=49315255, accessed November 5, 2009.

in Palo Alto, California.

By all accounts, the Datapoint engineers thought the article described a purely theoretical construction and paid no attention. And indeed, the networking system it described, Ethernet, relied on collision detection, a method they had examined and rejected.

But by the time that the article appeared, PARC had about 100 machines attached to the network, recalled one of the coauthors, Dr. Robert M. Metcalfe. Another research-driven project, the network consisted of a length of thick coax cable into which connections were made by drilling taps at intervals. (These were called "vampire taps.") This approach had also been examined and rejected by the Datapoint engineers, before turning to the previously described star and hub topology.

The original speed was 2.9 megabits, which was only slightly faster than ARC's 2.5 megabits. Metcalfe remembered[108] sitting in meetings where peopled argued that 1,200 bits per second was fast enough since, at that rate, the characters whizzed by on the screen too fast to read. Metcalfe noted that one reason of having the network was to let multiple office users share an expensive (and newly introduced) laser printer. To keep up with the printer's capacity of a page a second he figured the network would need a speed of 2.9 megabits.

Standard Ethernet speed was soon raised to 10 megabits, and in 1980 a consortium of DEC, Intel, and Xerox began promoting it, even before supporting hardware was readily available. Even the hiked speed of 10 megabits did not impress the Datapoint engineers, since they knew that a lot of the throughput was consumed by collision responses.

Wohl saw Ethernet in use for the first time at a computer convention in 1980. Later she was invited to see it in action at a Xerox training center in Virginia, and later at Xerox corporate headquarters in Connecticut. At the latter place, "I could see executives using networked Xerox workstations—which cost about $35,000 each—but I found that many of them had been carefully coached about what to say when an analyst came by and actually did not use them at all."

Having worked at Xerox to turn Ethernet into a product, Metcalfe left Xerox in 1979 and consulted for DEC, and then cofounded 3Com Corp, to make Ethernet interface cards. Later that year he was involved in an effort by the non-profit, standards-setting Institute of Electrical and Electronics Engineers (IEEE, pronounced Eye-Triple-E) to establish LAN standards. IEEE Committee 802 was set up for the purpose.

"I remember the day I made the call to Vic Poor, in about 1980, to say to him that we had this 802 committee making standards for local area networks. You have the world's leading LAN, so would you like to submit its specifications as a standard? He said he would get back to me in a couple of weeks, and as promised he did call back and say that Datapoint had decided not to submit to 802. So 802 went on to standardize three other LANs.[109]

"Not that ARC was not a beautiful thing, but they chose to keep it to themselves. That doomed them. Those were the years when open standards were not the thing, when firms

---

108 Metcalfe was interviewed by the author in 2007.
109 The three were Ethernet, IBM Token Ring, and (largely for industrial automation) Token Bus.

had proprietary systems. They were stuck in the IBM SNA[110] era. But since that time the trend has been towards open standards. Forgetting the technicalities, that business model was fatal to ARC," Metcalfe said.

At Datapoint, Jonathan Schmidt remembered that Poor wanted to go public, but O'Kelley opposed the idea, preferring to keep the ARC details proprietary. "O'Kelly wanted Datapoint to take over the world. It's the old story of a company that has some temporary success and some guy thinks that it's all his own work," he said.

With several LAN standards jockeying for market share, the race to become the dominant standard was on. Metcalfe sold his first Ethernet card, for the DEC VAX super-minicomputer, in 1981 for $5,000, running at 10 megabits.

ARC initially had a bigger installed base then Ethernet. In 1984, Datapoint was able to claim 5,000 ARC installations, while Ethernet could claim no more than 600 installations.[111] Datapoint also brought out an intelligent hub that year for network monitoring and control, ten years before anything similar was available with Ethernet.

On the whole, ARC was less expensive, but had the reputation of being slower, Metcalfe recalled. But, mostly ARC simply lacked the promotional backing that Ethernet enjoyed. "They (Datapoint) finally got second sources, but it was too late—Ethernet achieved escape velocity before they saw it," Metcalfe said.

But Datapoint was not exactly asleep. Datapoint published the ARC network specifications in 1983 in its "ARCNET Designer's Handbook" so that other vendors could make ARC circuitry. (One had already been doing so for two years, Fischer notes.) This led to the use of the term ARCNET (or ARCnet) to refer to the local area networking scheme used by ARC, while ARC continued to refer to integrated Datapoint systems based on ARCNET.

Using the resulting standard ARCNET chips, several suppliers brought out local network interface boards that fit into the IBM PC and used Novell Netware interface software, laying the foundation of the PC networking industry. Sales of ARCNET chips exceeded 300,000 in 1985 and 800,000 in 1988 Novell Netware resellers liked ARCNET because its was reliable and easy to set up, it performed better than Ethernet on networks with more than five machines, the hardware was cheaper, and the cabling (being thinner, and laid in short segments between hubs) was easier to plan, install, and maintain.[112]

The ARCNET Trade Association was formed in August 1987 to promote ARCNET, and apparently had some success in expanding the market. However, its efforts to remove ARCNET's stigma of being a one-vendor network by getting it recognized as an industrial standard by the American National Standards Institute (ANSI) hit various delays. By the time ratification came in 1992 there were various low-cost Asian clones of ARCNET hardware in the field that were not fully compatible.

Meanwhile, efforts were launched at Datapoint to enhance ARCNET, leading in September

---

110 SNA (Systems Network Architecture) and was a proprietary IBM networking system, introduced in 1974, for attaching terminals to mainframes.
111 MIS Week, March 14, 1984.
112 Michael Fischer, personal communication. He was the product architect at Datapoint in the late 1980s.

1989 to the announcement of ARCNETplus, which ran at 20 megabits per second, or eight times faster than the original version. (At this time, Ethernet was still restricted to 10 megabits.) Fischer recalled that it was the first local area network with a dynamically variable data rate, meaning that individual ports could operate at either 2.5 or 20 megabits. This made ARCNETplus compatible with first-generation ARC systems, while ARCNETplus ports could identify each other and operate at their full 20-megabit speed. Also, the maximum number of directly addressable machines on a network was raised to 2,047. But the presence of Asian ARCNET clones that were not fully compatible forced delays while a solution was sought. (The problem would not have occurred if ARCNET had been standardized in the early 1980s.) ARCNETplus was slow getting to market anyway, and Datapoint could not or did not pay for adequate promotion, recalled Fischer. So 20-megabit ARCNETplus was not able to make any market progress before 100-megabit Ethernet arrived in about 1993.

Triumph of the Pessimists

In 1989, ARCNET represented about 30 percent of existing LAN installations, but only about 17 percent of new LAN sales. Ethernet was already far outselling it, with projected 1990 sales of 2.3 million nodes, versus 1.6 million Token Ring nodes and 895,000 ARCNET nodes. ARCNET was much cheaper than the alternatives: $150 per connection for ARCNET, versus $413 for Ethernet and $761 for Token Ring. About 70 vendors were making ARCNET hardware. Pundits acknowledged that a 20-megabit version of ARCNET was coming out. But they saw little likelihood that it would succeed, since it was not an industry standard like Ethernet and Token Ring, and since there were no big computer companies pushing it. Ethernet had DEC, and Token Ring had IBM.[113] ARCNET and ARCNETplus had Datapoint pushing them, but by then, evidently, Datapoint just didn't count.

Those pessimistic pundits proved to be correct. Two decades later, in 2009, Ethernet could claim essentially 100 percent of the LAN market. Any LAN hardware for sale in any computer store was based on Ethernet. Circuitry equivalent to what 3Com sold in 1981 for $5,000 was available wholesale for about $1. Ethernet interfaces were built into PCs by default, and showed up in odd places like fire alarms.

Looking back, John Murphy felt that the difference between Ethernet's fate and ARCNET's fate was raw marketing clout. "I remember people noting that the amount DEC, Intel, and Xerox were spending to advertise Ethernet before Ethernet even existed was more than Datapoint's entire marketing budget," he said. "Datapoint, meanwhile, was not in the business of promoting a network."

Indeed, there were tales of Datapoint customers swayed by all the Ethernet advertising and inquiring if they too could get a LAN—and were surprised to learn that their ARC was a LAN.

But before becoming ubiquitous Ethernet had to evolve—and become pretty much like ARC. Ethernet ended up adopting ARC's star topology, where transmission is point to point (as from a computer to a port on a hub, and then out another port on that hub to another computer or to a port on another hub.) Therefore, an individual link in the network is used

---

113   The information in this paragraph is from the December 4, 1989, issue of Network World, page 25, article titled "ARCNET quietly takes its place in history." Michael Fischer, however, recalled that all published ARCNET sales numbers that he encountered, that he had any insight into, were low.

by only one machine, rather than having all machines on the network transmit into the same length of cable. This makes collision detection (and its impact on throughput) irrelevant, and data is stored and routed inside the hubs by local intelligence. The necessary cables are also thinner, hence cheaper, and easier to install. With evolved electronics, throughput rose to 100 megabits after 1993. (Speeds of one gigabit and 10 gigabits are now also available.) Meanwhile, the level of standardization that allowed network hardware from multiple vendors to be used in the same system was not available until the early 1990s—but that had been a feature of ARCNET from the start. Meanwhile, ARCNET could, and did, operate with the same signaling scheme over coax, phone cables, optical fiber, open-air microwave and optical links, and even (as a stunt) coat hangers. That's still not the case with Ethernet.

Metcalfe estimated that 250 million Ethernet nodes were being sold yearly. The ARCNET Trade Association noted that ARCNET remained popular in industrial and building automation and similar embedded applications where the network is not visible to the user. It estimated that a total of 11 million ARCNET nodes had ever been sold.[114]

---

114 See http://www.arcnet.com/abtarc.htm, accessed November 5, 2009.

Chapter 14

The Salad Days: 1977-1981

In 1977, Datapoint[115] became the $100 million company that its CEO, Harold E. O'Kelley, wanted to create. At the end of July that year it posted yearly revenue of $103 million, representing 43 percent growth over 1976. More importantly, 1976 had represented a 54 percent increase over 1975, while 1975 had represented 38 percent growth over 1974, and 1974 had represented an 83 percent growth over 1973, which had been the first year that Datapoint had made an annual profit. In other words, the average annual growth rate since 1973 had been a jaw-dropping 54 percent.

Speaking of profits, those amounted to $11.5 million for fiscal 1977, which represented a 47 percent increase over 1976. Previous year's growth rates for earnings had been 70 percent, 35 percent, and 75 percent. So, for earnings, the annual growth rate since 1973 had averaged 57 percent.

It was no wonder they started calling Datapoint a "darling of Wall Street."

For investors, these were addictive numbers—and indeed, when they stopped coming, the withdrawal symptoms nearly proved fatal. But at the time the uppermost question was not how to limit the damage when the growth inevitably stopped, but how to keep stoking growth. This involved introducing more and more products and entering more and more markets. But these efforts required continuous, expensive, on-going product development programs. Yet, as long as they generated growth, the development programs could be paid for, thanks to the magic of Wall Street.

For instance, in 1978 a new issue of 700,000 Datapoint shares was snapped up at $40 each. So the company raised $28 million purely from its reputation for growth. Operationally, that year Datapoint showed a comparatively puny profit of $15.3 million. The lesson: Wall Street magic was almost twice as potent as profits. If Datapoint's managers could show revenue growth, they could always conjure more money from Wall Street.

The previously cited 1981 "vanity" history of Datapoint included the following sentiment, apparently from Harold E. O'Kelley, referring to revenue: "Positive gradients showed results. As long as the gradients were sufficiently positive, any level of growth could be financed. It would be easy to raise all the working capital the company needed."

The words can be read as a pithy summation of a wildly successful corporate strategy, or as a justification for surrendering to an addiction. Applying the judgment of history, through 1981 the former was the case. After that, the addiction triggered a disaster.

But in the meantime Datapoint achieved some phenomenal accomplishments. Basically, between 1977 and 1981 it expanded its horizons from data processing to office automation and then to telephony, offering a continuous parade of new products or enhancements.

---

115 The financial and product information in this chapter was culled from various press releases, news stories, and Datapoint annual reports.

The Triumphal Procession

Inventing the local area network was not the only thing Datapoint accomplished in 1977. For the first time it bought another company: Amcomp, a maker of magnetic media, paying $2 million. Gone were the days when CTC/Datapoint was desperately trying to make ends meet by selling itself. (Amcomp owned patents in the disk storage field and began developing storage products for Datapoint. Results were mixed and Datapoint resold the operation in 1984.[116])

Michael Faherty also left the company that year, there being no further need for his skills as a turn-around artist.

Datapoint launched an entirely new product line, the Infoswitch division, selling computerized telephone management products. Datapoint's telecommunications management products (based on its processors paired with telephone switching circuitry) would eventually included automatic call distributors (ACDs) for routing incoming calls to call-taking agents, long-distance control systems for managing outbound long-distance calls, and various systems for logging calls and charging the costs to internal accounts. (Long distance calls were then a major source of corporate expenses.)

Datapoint also brought out the final expression of the 2200 line, the 6600. That machine used the same enclosure as the 2200 with its half-height screen (as did the earlier 5500) but used an enhanced CPU and the latest 16K RAM chips. A 6600 with a full load of 256K of RAM was nicknamed a Godzilla machine and supported 24 terminals. All later Datapoint machines would have full-height screens or none at all. (Datapoint processors are compared in Chapter 17.)

In October Datapoint also brought out the 1500, a self-contained desktop computer based on the Z80 processor recently marketed by Zilog, Faggin's startup. It was intended as a stand-alone unit for branch offices, and in hindsight looked enticingly like a personal computer. However, it was never marketed that way, and its $5,950 price placed it beyond the reach of that market.

Product development continued, but 1978 was memorable mostly for what happened in the federal courthouse in San Antonio. If Datapoint seemed to be living a charmed life, what happened there in June did nothing to change the perception, as Datapoint resoundingly won a case against the U.S. Government's much-feared Equal Employment Opportunity Commission (EEOC.) The EEOC had only recently been given the power to bring lawsuits, and chose to back Helen Sierra, who was severed from CTC during its financial crisis in 1970 after she got upset over the paltry nature of her raise. The company, complained the resulting lawsuit, discriminated against those with Spanish surnames in dismissal proceedings, hiring practices, and salaries. Apparently the EEOC did not realize that Datapoint was located in a city where Spanish surnames predominated—any firm that discriminated as described would have had a hard time hiring a staff, assuming it managed to stay in business. At one point the EEOC was demanding that Datapoint have 30 percent of its engineers be minorities within three years, although only 3 percent of engineers, nationally, were minority members. Datapoint refused to settle.

---

116   See http://chmhdd.wetpaint.com/page/Data+Disc+%28Datapoint%29.

The EEOC was not only unable to prove its allegations, but on the day the trial opened the first thing it did was to try to withdraw two-thirds of its exhibits, having decided they were invalid. (Perhaps, while driving to the courthouse after arriving in San Antonio, the EEOC lawyers noticed all the Spanish-language billboards.) The judge eventually ruled that the EEOC's case was "frivolous, groundless, brought vexatiously, unreasonably, and in bad faith," and awarded Datapoint court costs of $21,350 and attorney fees of $66,540, plus interest.

Datapoint was hailed as the David that slew Goliath and O'Kelley became a hero among his fellow CEOs. Datapoint executives went on speaking tours to talk about the case.

Meanwhile, Datapoint had revenue for fiscal 1978 of $162.3 million, up 57.5 percent. Profits were (as previously mentioned) $15.3 million, up 33 percent. The company had 3,889 employees.

In 1979, Datapoint brought out a full-scale office automation system based on the ARC. Called the Integrated Electronic Office System (IEOS) users of Datapoint desktop computers[117] on an ARC had access to central, shared files, and could communicate with each other via corporate e-mail, with file attachments. (E-mail to and from addresses outside the organization was not an issue yet.) The word processing was not the what-you-see-is-what-you-get graphical approach that would become familiar in later decades. As was standard at the time, it was based on typewriter emulation, with fixed letter spacing. Typographic features were denoted with special codes. Any cost-effective processor of the era would have had difficulty running graphics-based word processing software at any acceptable speed.

To extend the ARC from one building to another without having to string coaxial cable, Datapoint brought out the LightLink transceiver, which used infrared light to replace a length of coax. The devices usually sat on the roofs of adjacent office buildings.

Datapoint fiscal 1979 results were revenue of $232.1 million, up 43 percent from 1978. Profits were $25.25 million, up 65 percent. The number of employees rose to 5,066.

In 1980, among other things, Datapoint enhanced the IEOS with a facility called AIM (Associative Indexing Method), which automatically indexed the contents of the user's files to facilitate searching. Microsoft Widows 7 was hailed for including a comparable search facility in 2009.[118]

But the big news in 1980 was the introduction of a new operating system called the Resource Management System (RMS), plus the 8800 processor.

RMS was built around networking, as opposed to being a desktop system with retrofitted networking facilities (as remains the norm today.) Network security was no longer an add-on feature, as multiple passwords could be used to control access to any resource. Applications could be run from unintelligent terminals. An application that needed more RAM than its processor offered could go out on the network and claim unused RAM on another processor.

RMS would out-live Datapoint.

---

117 IEOS relied on Datapoint processors with full-height screens, especially the 3800 and later the 8600 processors. See Chapter 17 for a summation of Datapoint processors.
118 Michael Fischer recalled that AIM worked so well that no one bothered to write a stand-alone database program for the Datapoint environment, which made Datapoint look bad when compared to the PC environment of the later 1980s.

The 8800, meanwhile, differed from previous Datapoint processors in that it did not have a built-in screen, requiring that a terminal be attached. It resembled a small washing machine, and supported a (then) amazing one megabyte of RAM and a gigabyte of disk storage. (Datapoint processors are summarized in Chapter 17.)

That year Datapoint also expanded to Forth Worth, building an assembly plant there. The results of the 1980 fiscal year were revenue of $318.8 million, up 37 percent from 1979. Profits were $33.5 million, up 33 percent. The number of employees rose to 5,939.

More importantly, by the end of 1980 a $1,000 investment in Datapoint made in 1969 would have been worth $11,925, meaning the stock price had increased at a compounded annual growth rate of 22.9 percent.[119]

In April 1981, Datapoint set out to expand the concept of office automation by introducing the ISX, a digital PBX[120] that was intended to run in parallel with an ARC. Up to 20,000 users could have an ordinary analog telephone, a digital phone, or a computer terminal connected to a processor on the ARC. It was years ahead of any other PBX on the market, and probably the only one at the time that was trying to merge the worlds of office data and voice communications. The hugely hyped announcement was broadcast via satellite to 23 sites across the country from the Essex House in New York City.[121] Datapoint also brought out a compatible key system[122] called the KSX, which was resold from another vendor.

Later that year, in October, Datapoint held a ceremonial groundbreaking for a new $50 million Datapoint headquarter complex several miles north of its existing headquarters, on a 148-acre site on Interstate Highway 10 between DeZavala and Hausman roads. The first phase, with 400,000 square feet, was to be finished in late 1983, housing engineering, research, development, and customer service. The second phase, to start in mid 1982 and finish in early 1984, was to be a 10-story corporate tower and a four-story marketing, dining, and training center, totaling about 600,000 square feet. The third phase would include manufacturing and warehouse space, to be finished early 1985.[123] At the time of the announcement, the firm's 3,400 employees who worked in San Antonio were housed in 36 different buildings.

Other product announcements that year included the 8600 desktop processor, intended for IEOS use. It had a detachable keyboard, an amber screen that was supposed to produce less eye strain than a traditional green screen, 256K of RAM, and an integrated ARC interface. (Datapoint's processors are compared in Chapter 17.)

Datapoint also introduced a 20 page-per-minute 9660 Laser Printer, costing $65,000. At the time there were no more than two other laser printers on the market. Based on a Minolta laser printer engine, the 9660 offered features that were years ahead of its time, including workgroup sharing, queuing, job prioritization, remote monitoring, local font storage,

---

119 Untitled stock analysis in the Datapoint vertical file of the San Antonio Public Library.
120 A PBX (private branch exchange) connects the phone extensions used inside an office or enterprise to each other and to the public switched telephone network (PSTN.) The PBX may make the connection automatically, but at the time many PBXes used an operator sitting at a switchboard.
121 As will be shown, the ISX would become a cruel disappointment for Datapoint..
122 A key system is a set of telephones intended for a smaller office, where the users select the desired line by pressing a button.
123 None of it was ever built.

lockable input and output trays for sensitive financial and accounting tasks, and of course easy integration with an ARC network.

Datapoint also brought out a fax interface (later a standard feature in PC modems.)

By the end of the year Datapoint also unveiled its CBG (Color Business Graphics) system, costing $30,000 for the processor and digitizing tablet, $15,000 for a color printer, and $10,000 for a 35mm slide recorder.

The CBG system was developed by a team led by Herb Baskin, the UC Berkeley professor who had been consulting for Datapoint in Silicon Valley. In 1976 Datapoint had bought his firm and turned it into the Datapoint Western Development Center, with Baskin as vice president. Other accomplishments of the development center included a version of COBOL that met strict U.S. Government standards. (Baskin would quit UC Berkeley in 1980, and then Datapoint in 1981, for a new startup.)

The results for fiscal 1981 were revenue of $449.49 million,[124] up 41 percent from fiscal 1980. Profits were $48.761 million, up 46 percent. The number of employees had risen to 7,915. Datapoint had facilities of various kinds in San Antonio, Waco, Austin, Odessa, and Fort Worth, Texas, and in Sunnyvale, California, and Toronto, Canada.

Since 1977, Datapoint's average annual revenue growth rate had been 44 percent. The average annual earnings growth had been a nearly identical 45 percent. Overall, the company grew 336 percent during that period in terms of revenue.

The growth continued. The procession of new products continued.

What could go wrong?

Breaking Free from TRW

One problem looming on the horizon concerned the TRW deal. As previously described, CTC was in desperate circumstances in 1970 when it gave TRW marketing rights for its products overseas. Partially renegotiating this contract was one of the first things that Harold E. O'Kelley did when he became CEO in 1973. Now that he was running a large, successful firm he saw no reason to let it remain hobbled by an old agreement that was made under duress. And as 1980 rolled around, there came the realization that the agreement would expire in 1983, and would either need to be renewed, or be replaced with some other marketing arrangement. Otherwise, after 1983 Datapoint would have no overseas sales.

"The agreement was a wonderful thing for Datapoint in 1973 since we could barely market domestically, let alone internationally," Ed Gistaro recalled. "But due to our domestic success, O'Kelley started feeling his oats. On top of that, we had no exit strategy. We had no capability for taking over international sales by 1983, and in 1980 international sales amounted to 40 percent of our business. Sales were scattered over 27 different countries. What were we going to do in 1983? O'Kelley did not want to have to go back to TRW and make another deal.

---

124 Despite the many events cited in the next chapter, these figures never officially restated.

"So here I was running marketing, happy as a clam, and O'Kelley said, 'Ed, I'm going to pull you out of there and I am going to put you in charge of corporate development. At least, that is what we're going to tell the world you're doing, but you're going to be doing one thing—get us out of that contract.' I said, 'How are we going to be able to do that, Harold? You've been trying to do that for the last four years, and you haven't been able to do it.' He said he thought his personality was getting in the way.

"He said, 'I don't care how you do it, but do it.' So I started meeting with the right-hand man of the head of TRW. At this point international sales were a big part of Datapoint, and Datapoint had become a big part of TRW. They were making a lot of money, and yet they did not have to manufacture or inventory anything. They were getting the lion's share of the revenue. They were buying stuff from us at very low prices and reselling it overseas. They were getting all the sales and training revenue. I was getting nowhere talking with them.

"I said, 'Bob, we're going to take this thing back anyway in 1983.' He said, 'How are you going to do that?' And I thought, you know, he's right, we can't do that.

"But there was a company in New England called Inforex. They made key entry[125] stuff, and the stuff that we and our competitors made was replacing it. Inforex happened to be run by a guy who used to be my old boss at Honeywell. He called me to say that they were going bankrupt but they might have some assets I might want. A light went on.

"I asked, 'How are you set up internationally?' He said they are really strong in Europe with offices in all the major countries, and all the major cities. So I wondered what would happen if I took over this company for the express purpose of training all those people on Datapoint products?"

Datapoint eventually paid about $15 million for the company.[126]

"The guys at TRW went ballistic, saying, 'Ed, you can't do this, you can't train these people on your equipment,'" Gistaro recalled. "I said, 'Why not?' They said, 'You have to stay with us.' But I said that we are staying with you. What's the problem? They said, 'You don't really think you can all of a sudden turn this whole thing over to these guys in 1983, do you?' I said, 'Yes, I think we can.'

"Now, could we have done it? I kind of doubt it. But did he think that I was nuts enough to think we could? Yeah. So they said, 'Okay Ed, you can have the whole ball of wax for $100 million—wholly owned subsidiaries in 16 countries and distributor relationships in another 11 countries.' We paid another $17 million for the projected profits they were going to make between 1981 and 1983. That will give you an idea how big that operation was."

Datapoint was able to raise the money through Kidder Peabody and simply wrote a check to TRW.

"Some people thought I was nuts because there were almost no assets involved in the sale, but the assets were the ability to sell products overseas. I remember calculating that the $117 million let us generate, over the next few years, more than $2 billion in revenue,"

---

125 Key entry systems, or key-to-disk systems, were used to enter data from a keyboard and store it on a disk, from where it would be later loaded onto a tape for mainframe use.
126 Datapoint resold Inforex in 1985.

Gistaro said.

Herb Baskin, Datapoint's software developer in Silicon Valley, remembered welcoming the news. "Datapoint had become a de facto prisoner of TRW," he recalled. "Representing 40 percent of sales, they would steer the Datapoint ship a good deal of the time. It distorted Datapoint's trajectory in a way that was more profound that anything I can convey. Datapoint had a better feel for the future development of their products, while TRW just wanted to follow their customers' demands, plus they had a different class of customers. There was animosity between the two companies that was self-defeating."

But, thanks to the happy coincidence of a single phone call, the situation went away. Datapoint was able to eliminate the middleman and take direct ownership of the overseas distribution network for Datapoint products that had previously belonged to TRW.

Obviously, it was another success story for a company that seemed fated for success. But Gistaro would later see it as the beginning of the end.

Xanadu in San Antonio

Among other things, Datapoint was now a company that could afford luxuries. One such luxury was the hiring of futurist Ted Nelson,[127] who is generally credited as the inventor of hypertext. He was in the midst of a life-long quest to produce a global multi-user hypertext-based system called Project Xanadu. The World Wide Web (which also relies on hypertext) would later superficially resemble his earlier conception of Xanadu, but Xanadu would include strict version control for documents, and two-way links. (On the World Wide Web you can link to a document, but there is no link back from that document.)

In mid-1981 Nelson was editor of Creative Computing magazine and living in Pennsylvania. "Circulation had increased under my watch but the overall enterprise had suffered due to their very bad software division, and there was a pitch for me to come down (to Texas) and help design new office software.

"'Well, you want to move to Texas?' I was asked in the interview, and I was sure I wasn't but it was the best thing that I had going. I was later surprised to find that Texas was very mellow," he recalled.

Nelson (offspring of a Hollywood director and a movie star) drove to Texas in a Dodge van jammed with his possessions and his personal programmer. The latter was using an Apple II to work on Xanadu code as well Nelson's personal word processor, Juggler Of Text (JOT). The two projects were excluded from his Datapoint employment contract. The drive took more than 24 hours.

He was inside the city limits of San Antonio when a policeman pulled him over.

"He was the meanest-looking son-of-a-bitch policeman you ever saw—tall and rangy and anorexic, with a SWAT badge. He was a caricature of the meanest Texas state trooper you ever saw.

---

127 Nelson was interviewed by the author in 2009, and on previous occasions.

"'Officer, am I under suspicion?' I asked jovially. He said that someone had been stealing TVs from condos in the area. I nearly laughed at the idea of driving all the way to Texas to steal TVs. If I had had any alternative I would have turned around and left, but luckily I did not. But it was so ironic that my first experience with Texas would be a ghastly validation of every stereotype," Nelson said.

Later, he showed a Datapoint research executive the text processor on the basis of which he had been hired. "He was horrified that you could not move the cursor. Instead you stepped between words, sentences, and paragraphs. It was based on my previous design for Xanadu, where you never move text and instead appended changes and maintained a list of current contents, much as Wikipedia does now," Nelson said.

Computer interfaces did not represent a new topic for him. "I had visited PARC[128] in 1975, I had seen Engelbart's[129] lab in 1967 and saw the mouse. On that day I decided the mouse and not the light pen would be the pointing device of choice. Also, that skateboards would be the transport of choice, since they go well between walking and bikes."

He worked on a new Datapoint word processor from October 1981 into January 1982, by which time the design was complete and he was waiting for programming to start.

"I contend that interface design is a branch of cinema—it's presentation art. We did a number of demos in the beginning to emphasize that we were doing something different," he recalled. These included eye-popping ways of suddenly painting the text on the screen, rather than scrolling it down as it arrived.

A manager mocked the special effects as operator entertainment, unnecessary for people who were doing their jobs. In response, Nelson took a couple of executives to an arcade, and found they had never seen a video game.

"Vic Poor got it, although he was grouchy about it," Nelson recalled. "He was behind my idea and had the political savvy to tell me to not show my documents to any of the old IEOS people."

The project was killed anyway. "It was part of some larger package, and was shot down as too futuristic—they had not seen what was going on elsewhere," Nelson said. "I was extremely depressed. They had a version of Pac Man for Datapoint machines, and I played that for a couple of months." He was later given the job of writing documentation for RMS, and was laid off after about three years with Datapoint.

Supported by a book deal, he stayed in San Antonio until 1987, after Autodesk Inc. began funding Project Xanadu,[130] and then moved to California.

"Datapoint was on the cusp between the mainframe and the personal era—if they had played their cards right, dropped the price, and marketed to the personal market they could have done well. They would have beat hell out of the IBM Personal Computer since the PC didn't have Datapoint's full package of software.

---
128 The Xerox Palo Alto Research Center, where Ethernet was invented and the graphical user interface was pioneered.
129 Douglas C. Engelbart of the Stanford Research Institute, who invented the computer mouse that year.
130 Autodesk dropped the project in 1992.

"I liked the people but they were so clueless. I was telling them that the personal computer market was coming and they would have to compete with it. People always called me crazy, although I'm used to that.

"But Datapoint was a decent, innocent company trying to do old stuff, unlike Microsoft, which was (having forcible intercourse) with the future.

"Datapoint's legacy is that they put this good-looking thing on the desk, with a clean operating system that's preferable to what a lot of people have to deal with today," Nelson concluded.

Chapter 15

The Debacle: 1982-1985

February 1, 1982 was a Monday. Datapoint's stock closed at $49.38 per share. A total of 97,000 shares traded—a normal amount on a normal day.

It was the last such normal day that Datapoint would ever know.

Not that there wasn't trouble in the air already. Basically, the early 1980s were a wrenching time, economically, with recessions and recoveries following in quick succession. The gross domestic product (GDP) plunged a whopping 7.6 percent in the second quarter of 1980. It fell only another 0.7 percent in the third quarter, and then there was a recovery during the next two quarters, with GDP growth of 7.6 and 8.4 percent. Then the GDP fell again in the second quarter of 1981, by 3.1 percent, only to rise again in the third quarter, by 4.9 percent. However, things got even worse, with another decline of 4.9 percent in the fourth quarter of 1981, followed by a further plunge of 6.4 percent in the first quarter of 1982.

Meanwhile, the prime interest rate was rising to unprecedented heights, surging past 20 percent several times during this period. In February 1982—the month in question—it still hovered around 17 percent.[131]

Obviously, this was not a good time to be selling computers (or anything else.) Obviously, the sales force could not be expected to continue making the same kinds of numbers they had brought in during the recession-free years of the second half of the 1970s. Then, Datapoint's annual growth had averaged an eye-popping 47 percent for both revenue and earnings.

Except that (as events were to demonstrate) the sales force was indeed expected to continue making sales as if conditions had not changed. Salespeople faced relentless pressure to meet their sales goals, or lose their jobs. As a Datapoint sales trainer was heard to say to a class of new salespeople, "This company asks, 'What have you done for us TODAY?'"

That unbending attitude was not simply the product of arbitrary greed—Datapoint had a reputation to uphold, which required making enough sales to maintain an impressive growth gradient. Meanwhile, Datapoint executives and spokespeople made the mantra-like boast of "X quarters of uninterrupted growth" a part of every presentation and statement. At that time, X had reached 39, and apparently no one wanted to be the first to say that there would be no number 40.

Maintaining the gradient and the unbroken string of growth quarters was important because Datapoint had a full plate of development projects underway (as explained in the previous chapter.) But what if growth came to an end?

For a warning, they only had to look at their competitor Data General. In September 1979, after 11 years of continuous quarterly growth, it announced a down quarter. The stock price

---

[131] The figures are from various tables available on the Web site of the Bureau of Economic Analysis of the U.S. Department of Commerce, www.bea.gov, accessed March 2, 2009.

promptly fell 20 points.[132]

Meanwhile, besides the stock price and the revenue growth gradient that sustained it, there was a third, more arcane factor at work: the price-earnings (P/E) ratio. Stock analysts use the P/E ratio to gauge whether a stock is a solid investment or an over-hyped speculation. The higher the P/E ratio, the more likely it is that the price is based on hype—unless the stock actually lives up to the hype by producing substantial growth. Datapoint had been living up to the hype, and (as events would demonstrate) it had been bought by a lot of institutional investors. Unfortunately, those were the people who could unload their stock at a keystroke.

Three years earlier, in 1979, a Forbes magazine writer thought it noteworthy that Datapoint's P/E ratio was as high as 15.[133] But the ratio had since zoomed to 27.6, based on the most recent fiscal yearly earnings of $2.45 per share and a stock price that had peaked at $67.50. Historically, the average P/E ratio, across all eras and industries, going back to 1881, was about 16. But in early 1982, the battered economy and soaring interest rates had driven down stock prices, and the average P/E ratio hovered a little above 7.[134] So Datapoint's stock stuck out like a sore thumb, and only its history of growth made it acceptable in the portfolios of professional money managers.[135]

Also, amplifying what was about to happen, was a Wall Street rule of thumb that bad news is always followed by more bad news. There is more to this attitude than simple cynicism—corporations spend a lot of effort maintaining upbeat facades. They don't drop those facades and admit to bad news for trivial reasons. Any admission of a problem usually represents the tip of a bad-news iceberg.

Within the computer industry there was, at the time, an additional layer of negative expectations. There was a wide-spread (but, as it turned out, mistaken) assumption that the number of computer companies had grown too fast and the market had gotten too crowded. Soon (said the pundits) the market must suffer another shakeout, similar to the one that hit the mainframe vendors in 1970. Speculating on which computer vendor would be the first to head for the exit was a favorite game among industry observers.[136]

In hindsight, it seems fair to say that Datapoint's financial position in early 1982 was a product of the push generated by its growth, and the pull generated by skepticism over its P/E ratio and the expectation of a shakeout. In January 1982, those two forces were nearly balanced.

The event that would drive them out of balance was already brewing in the Datapoint sales department.

---

132 Forbes, December 10, 1979.
133 Ibid.
134 Both the average P/E ratio and the 1982 figure are derived from data used in "Irrational Exuberance," second edition, Robert J. Shiller, Doubleday Business, 2006, posted at www.econ.yale.edu/~shiller/data/ie_data.xls.
135 Gistaro recalled in 2009 that, at the time, the rule of thumb was that a corporation's P/E ratio should equal its annual earnings growth rate. In early 1982 its P/E ratio, nearly 28, was actually well below its fiscal 1981 earnings growth rate of 46 percent.
136 At a computer trade show in 1983, the author and a companion got so sick of hearing shakeout speculation that they nearly talked themselves into running through the aisles shouting, "Shakeout! Shakeout! Run for your lives!" It would have been an interesting experiment.

Sunshine Pumps

Actually, managing any industrial sales force is famously difficult. Former Datapoint executive Gerry Cullen recalled[137] being a temporary vice president of sales during this period while an executive search firm located a "permanent" vice president. Piles of beautifully formatted sales projections arrived from the sales offices at prescribed intervals, but he gradually became convinced that they were simply pumped up with random numbers. Mismatches between projections and reality—when they were even acknowledged—were shrugged off as complications caused by the resellers. His attempts to meet with actual customers or prospects to confirm selected projections were stonewalled by the regional offices. In six months he was only able to meet 18 customers or prospects. As a result of his efforts the sales force started calling him The Mole and were complaining to higher management.[138]

Yet, the beautifully formatted sales projections were the basis of factory production planning, and triggered large investments in raw materials and labor.

He consulted a friend at another company, who had field sales experience. The friend told him to stop making trouble—as long as sales were going up, the projections didn't matter anyway. All anyone would care about was that the printouts were impressively thick, so that what his friend called the "sunshine pump" could continue functioning. And, indeed, after he started nosing around, the submissions doubled in thickness, Cullen recalled.

Looking further into the situation, he found that the average tenure of a vice president of sales, industry-wide, was about 18 months. After the firing of the previous incumbent because of sagging sales, the new person would have six months to get situated, and then would usually spend six months identifying and firing the worst salespeople, and then would spend the next six months finding replacements. At the end of that 18 months sales would be no better, so the new person would be fired in turn (often with six months' severance pay) and the cycle would renew.

Enjoy the ride, he was told.

Cullen recalled that Datapoint had nine vice presidents of sales in 12 years—which he saw as a commendable record. But, "My opinion persists today that the top management had no idea why the products were popular and, therefore, offered little help to the vice presidents of sales," he recalled.[139]

Shifting Numbers

Meanwhile, at the start of 1980, Ed Gistaro was still the head of marketing, where he had a specially picked financial controller for the sales department, tasked with riding herd on the sales force.

---

137 The observations are from "The Coldest Call," published in 2007 by Gerry Cullen, Austin, Texas.
138 Tourists from headquarters made the field sales force nervous anyway. According to Cullen, there had been an earlier incident when Datapoint was thrown out by a large prospective customer—a bank—after a visiting Datapoint vice president told them that their approach to data communications was "dumb." He then sketched out his own approach on a whiteboard. The customer became enraged, since the approach that the Datapoint vice president blithely described violated federal banking regulations
139 E-mail to the author, March 16, 2009.

"Our incentive plan put a lot of money on the table, and there was a huge incentive to cheat," Gistaro. "So we had to have really tight controls in marketing."

But in 1980 Gistaro was reassigned to negotiate the acquisition of TRW's international sales network, as described in the previous chapter. Decades later, Gistaro decided that the moment he took that transfer was the moment when Datapoint started running into trouble. Because when he left, Gistaro's handpicked controller was transferred to manufacturing. As for the new man, "He later left us and did the same crazy stuff at another firm and wound up in jail, as I understand it," Gistaro recalled.

After the acquisition of the TRW network was finished, Gistaro was put in charge of running the new overseas distribution network, becoming vice president of international operations. He remained unconnected with marketing.

He remembered having an ominous conversation during a golf game with another Datapoint executive during that period. "I asked how things were going, and he said, "Great. Remember that Chicago office? They weren't bringing in any sales, but we got things straightened out in two weeks and now they are beating their goals.' I pointed out that the (sales) lead times on our products were about six months. You can't turn something around it two weeks. He was very inexperienced when it came to marketing."

Then, in late 1981, Datapoint CEO Harold O'Kelley came to Gistaro with a problem. "He said, 'Ed, we've got some numbers coming in that I don't quite understand. Could you look into it? So he makes me the chief financial officer, and I'm no longer the international guy. I didn't have any formal financial training.[140] I said, 'Harold, don't you need a plaque or something on the wall saying you know something about accounting (to be CFO)?' But he said that I had just pulled off one of the most amazing financial transactions in the technology industry with the TRW thing. 'We have plenty of guys who can do accounting and bookkeeping. We need guys who can think, and know business strategy,' he said.

"I think that I was a reasonably decent CFO. I brought back the controller and told him that we think we have a problem. The reverse revenue numbers (the numbers that had puzzled O'Kelley, representing goods that had been sold by Datapoint but then returned) have been going up month by month for quite a while now, and nobody knows why," Gistaro recalled. Other executives in the sales department were given lateral transfers to give the returned controller a free hand.

The controller soon had results. "The words he used were, 'We have a conspiracy here—they are all in on this thing to keep reporting more and more revenue.' When I saw the numbers I went to Harold (O'Kelley, the CEO) and I said that we have a major problem, we have to disclose this. So we brought in independent auditors and an independent law firm. They wrote a huge report and pretty much nailed what had been happening," Gistaro recalled.

"Later, when things were going really bad, O'Kelley asked me why we didn't have these control problems when we were growing willy-nilly. I said, 'Harold, I won't say that I was never afraid to tell you bad news, but I did tell you the bad news. The people you have now are not only afraid, but they won't tell you the bad news. That is what happened.' I was,

---
140 Both Gistaro and O'Kelley had engineering rather than financial backgrounds. This would become a sore point with Wall Street analysts.

incidentally, afraid to tell him things, but it was a matter of trying to hide things, or telling him and taking my lumps. I was pretty sure he was not going to fire me."

David Monroe could not share that belief, as he encountered an entirely different attitude on the part of O'Kelley. By that period Datapoint had been divided into five divisions and Monroe was the head of one, the Office and Graphics Systems Division.

Along with another executive, he was additionally being groomed as a future CEO, perhaps five years down the line. The grooming included special training, during which he heard O'Kelley say that the company's biggest problem was to ship the products fast enough to meet demand, especially at the end of each quarter.

So he dutifully set about examining the logistical operations of his division to see what he could do to expedite end-of-quarter shipping. He noticed that they often shipped to the same customers at the end of each quarter, and that the hardware was often sent back during the next quarter.

He began to suspect that there was something wrong with some of the orders. In the end he went outside the chain of command, going over the head of his boss and taking the matter straight to O'Kelley.

This, he reflected more than a quarter century later, may have been a mistake.

O'Kelley listened to Monroe's concerns—and then, taking advantage of being taller than Monroe, stood up behind his desk and leaned forward with his knuckles on his desk until his face was in Monroe's face.

"You're not fired yet," he told Monroe, adding that the former technician would be better off returning to his oscilloscope, and minding his own business. They took his division away and busted him back to vice president of research and development.

Monroe had other options, though, and resigned to set up his own company with the intention of marketing the videoconferencing system that he and Frassanito had been experimenting with, as described in Chapter 11. But it did not turn out that way. As explained in Appendix A, the subsequent fate of his invention remained entangled with that of Datapoint—in the end, fatally so.

Groundhog Day

Exactly who knew what, when, was never established—and anyway was not a factor in the disaster that was about to engulf Datapoint. That was triggered by one tiny piece of information.

The problem was that Datapoint's management had previously taken the sales bookings at face value and estimated that the earnings for its second fiscal quarter, ending January 31, 1982, would be 66 cents per share. The company had made 60 cents per share in the same quarter of the previous year,[141] so the estimate represented an annual growth of 10

---

[141] To negate the effects of seasonal business cycles, quarterly comparisons are typically made between the same three months of consecutive years. In this case, the second quarter of fiscal 1982 was compared to the second quarter of fiscal 1981, not to the immediately preceding first quarter of fiscal 1982.

percent. This was well short of the 40-plus percent average growth rate that had previously mesmerized Wall Street. But, considering the economic climate, no one was particularly surprised. The stock price, however, began drifting downward, below $50.[142]

Then, on Groundhog Day, the company's fate literally turned on a dime.

Responding to the discoveries that Gistaro had made concerning fake sales, Datapoint issued revised quarterly figures on Tuesday, February 2, 1982. The new figures showed that the firm remained perfectly profitable. However, second quarter per-share earnings would be a dime less than previously estimated—56 cents instead of 66 cents per share. Since per-share earnings in the second quarter of the previous fiscal year had been 60 cents, the new figure represented an earnings decline rather than growth.

In other words, Datapoint's unbroken string of year-over-year growth quarters had ended.

It was as if some spell were broken.

Suddenly, there would be no 40th quarter of unbroken growth. Suddenly, Datapoint was no longer the darling of Wall Street—it was just a firm whose P/E ratio was unjustified. And, if there was going to be a shakeout, Datapoint suddenly looked like the company that was closest to the exit.

That day the stock price fell from $49.38 to $41.00, with 826,000 shares traded.

Wednesday, the price fell to $36.25, and 1,118,500 shares traded. Of about 19.5 million shares outstanding, about two million shares had traded hands in two days. Thursday saw a rebound to $37.25, and sales had slowed to 472,400 shares. Friday saw a further slump, to $34.88, with 364,900 shares trading hands.

The slide continued after the weekend, to $33.25, with 450,200 shares sold. At that point the price had lost a third of its value and the shareholders had lost $320 million.

Tuesday (February 9) arrived. A week had gone by since Groundhog Day, and it almost looked like the excitement might be waning. The price rebounded to $33.75, and 481,200 shares sold.

But that day syndicated financial columnist Dan Dorfman came out with a column blasting Datapoint. He accused Datapoint's management of dumping their stock before Groundhog Day, avoiding losses of $8 million. In response, O'Kelley insisted that he had sold some stock only to pay taxes, and it later came out that Datapoint insiders had bought more stock than they had sold during the previous 18 months. Nothing ever came of the accusation, but the next day (February 10) the slide resumed. The price reached $31.50, although "only" 389,200 shares were traded.

Thereafter the slide continued, day after day, as unfounded rumors swirled through the financial markets about what was really wrong at the company. Sales fell as customers retrenched in response to the falling economy, but also as they questioned the firm's prospects of survival.

---

[142] The situation, and Wall Street's reaction, was spelled out in a New York Times News Service story that appeared on February 19, 1982.

On February 18 Datapoint executives held a conference with some influential market analysts in New York City to try to reassure Wall Street. They only made things worse when they backtracked on a previous statement that the second half of the fiscal year would be better than the first half.

So the stock price continued falling, becoming an on-going joke and embarrassment. Previously, the company's stock price had been carried in the opening screen of its internal e-mail system. It became a source of anguish and was removed. A pair of vultures started nesting atop one of two adjacent 10-story buildings that Datapoint used as office space, located uphill from the factory it had built 12 years earlier. Perched on the edge of the roof, they seemed to study the employees as they walked to and from their cars in the parking lot, enhancing the gallows atmosphere.[143]

Datapoint, too, retrenched. By March 26 it had frozen wages and hiring, 230 people were laid off in San Antonio, a factory in Waco had been closed idling another 350 people, operations had been scaled back at Datapoint's development center in Berkeley, California, and construction of the planned new headquarters had been put on hold. News of each move triggered further bad publicity on Wall Street, fed the rumor mill, confirmed the expectation of further bad news, and further drove down the stock price.

On April 8 the stock price fell another $3.25, closing at $22.12, after the previous day's announcement that Datapoint would lay off 250 people from its sales force of 1,200 (in 60 offices). Also, it announced that results for the third quarter were expected to be "disappointing."

What had not been announced on Groundhog Day was that the missing dime in the second quarter per-share earnings came from orders that were previously on the books, but which were subsequently found to be bogus. Gistaro (who was made Datapoint's chief operating officer in March) mentioned this in a media interview published April 25, but there was no specific reaction.

Meanwhile, in early April he had also instituted a three-week "amnesty period" for the sales force. They were told they could remove from the books any sales they thought might be shaky, no questions asked. The results were horrifying, but the executives felt that it would be better to come out with the bad news all at once, and end the drip-drip-drip of negative reports that had been eroding the company's reputation like beach sand. However, huge sell-offs began as soon as rumors began to circulate that a loss would be announced.

First there was Black Friday on April 30, when the stock price fell $4.87 to $16.75, despite trading in the stock being suspended for more than three hours. After the weekend came Black Monday, when the price fell to the ill-omened value of $13.13, despite another temporary trading suspension. The volume that day was 1.2 million shares. To cap things off, the Wall Street Journal that day ran a news feature saying that Datapoint's value-added-resellers (VARs) had been pressured to write shaky orders to keep the sales numbers elevated.

After the market closed Wednesday, Datapoint's management made their announcement:

---

143 The facilities manager refused to shoot them, saying it would be illegal.

Datapoint had lost $22.9 million in its third fiscal quarter. Moreover, they had removed from their books new orders worth $105.9 million, suspecting they were bogus. But plenty of other new orders that were considered firm, worth $117 million, remained on the books.

And the next day (May 6) the stock price made headlines when it actually bounced back—to $13, rising $1.75. But in 95 calendar days Datapoint's stock had lost about three-fourths of its value. On paper, the stockholders had lost about $800 million

One other chore remained: the next week, on "Black Thursday," five executives (including three vice presidents) were fired. A sixth was demoted and later left.

There was to be one more humiliation. In the Wall Street Journal, 500 words is a major article. But on May 27, 1982, it ran an article nearly six times that length, starting on the front page, detailing marketing sins it had uncovered at Datapoint. Basically, it found, there had been a rush at the end of every fiscal quarter at Datapoint for the last couple of years to send out products to distributors, since revenue could be booked as soon as an item was shipped. They shipped to customers who had not met Datapoint's own credit requirements, and they shipped to customers who did not want the hardware and who would accept it only if Datapoint would pay warehousing fees. They even shipped to a "Joe Blow" in a hotel room in South Padre Island (a Texas coastal resort area) assuming that, by the time delivery efforts failed and the shipment found its way back, a real customer would be ready to take it. The distributors knew that when crates arrived stained with beer and taco sauce that the shipping crew had worked overtime at the end of the quarter, eating Mexican food that Datapoint paid for, to get everything possible shipped. There was so much unwanted Datapoint hardware floating around that distributors were brokering it to each other, which cut into the business of Datapoint's own salespeople—who were told to be quiet if they complained.

After that, the rain of bad news ended, and the situation began to stabilize. The economy also began to recover.

But that was not the end of the story.

"Secret Marketing Plan"

More details came to light four years later, after Datapoint's auditor, Peat Marwick Mitchell, sued Datapoint for breach of contract,[144] saying it was damaged by being dragged into the SEC investigation of what happened. (This was after Datapoint settled a resulting investors' class action suit out of court. Peat Marwick Mitchell did not participate in the settlement, and then sued Datapoint, saying Datapoint had lied to it.)

The suit said there had been a "secret marketing plan." Peat Marwick Mitchell alleged that, "Senior management forcibly encouraged and pressured individuals in the marketing division, despite the economic downturn, to take steps sufficient to achieve management's unbending goals and to take steps to reflect enhanced earnings. Senior management created an atmosphere in which the realization of their goals permeated everyday

---

144 Peat Marwick Mitchell v. Datapoint, whose file was examined by the author in the San Antonio federal courthouse in 1986.

operational decisions, and tacitly approved and/or recklessly perpetrated and accomplished the secret marketing plan."

It described five different types of misconduct, and said that the scheming began in February 1981—a full year before the debacle began. It said that the scheme began in Datapoint's eastern sales division, although Nevada and California customers appeared on the list of irregular sales.

First, it said that Datapoint had booked sales revenue from hardware shipped in advance of the agreed delivery date, in the absence of any customer agreement to accept early shipments. The suit listed numerous examples, with hardware being shipped as early as 10 months before the customer wanted it. In one case it arrived before ground had been broken for the building that was to house it.

Second, it said that Datapoint booked revenue on hardware shipped for orders that had already been canceled at the time of the shipment. Two examples were given, and one amounted to $700,000.

Third, it said that Datapoint actually booked revenue on hardware shipped when there was no sales agreement. It listed one example involving more than $400,000.

Fourth, it said that Datapoint booked revenue on partial shipments, which violated standard accounting rules. It listed two examples. One involved a reseller who agreed to take the partial order only if Datapoint would pay for warehousing and financing. Datapoint did not agree and shipped the items anyway. The second example involved, oddly enough, Peat Marwick Mitchell itself. In that case the partial shipment, worth $450,000, arrived almost a year early.

Fifth, it said that $12.3 million of revenue that arrived in the first two days of fiscal 1982 was credited to fiscal 1981.

All the irregular shipments were refused by the recipients and sent back, which is how they came to light. Apparently there were other irregular shipments that were accepted, and became "real" sales. There were reports that Datapoint would try to have salesmen present when irregular shipments arrived, to try to talk the customers into accepting them. To induce the customers to not send back the unwanted hardware, sales executives ended up paying warehousing fees out of office funds or from their own pockets.

The final tally was that revenue for fiscal 1981 was claimed to be $449,490,000, which was overstated by $22,100,000. The reported profit of $48,761,000 was overstated by $11,100,000 and a reported earnings per share ($2.45) was overstated by 25 cents.

The whole thing ended with a whimper. In June 1984, Datapoint as a corporation, and one individual who had been an executive in Datapoint's eastern sales region, consented to an order by a federal judge that they commit no further violations. And that was that. There were no criminal indictments. Nor was there any requirement that Datapoint restate its revenue and earnings, and the inflated figures continued to be used in the historical sections of subsequent Datapoint annual reports.

The reason why the government was so unexcited by the scandal—even though it wiped out nearly a billion dollars in investor wealth—presumably stemmed from the lack of "materiality." In auditing, a material change is a change in a firm's financial picture large enough that it might change someone's mind about investing in that firm. The rule of thumb is that, for revenue, a material change is a change exceeding 5 percent.

For the total revenue claimed by Datapoint that year, the $22.1 million overstatement amounted to "only" 4.9 percent.[145] In other words, despite all the excitement, the overstatement produced by the scandal wasn't even material. The "secret marketing plan" had an impact that was officially beneath notice. The fact that the plan destroyed the company's reputation on Wall Street (and that the results were apparently considered material by plenty of investors, who dumped the stock) was not any concern of the SEC's. (Recall, however, that another $105.9 million in doubtful orders was purged from the books in April 1982. Had those "sales" gone on to be shipped, booked as revenue, and then reversed, they would easily have had a material impact on subsequent quarters.)

The damage was done not in the law courts, where word games rule, but on Wall Street, where reputation is reality. On Wall Street, Datapoint had previously been able to live on its reputation for growth. On Groundhog Day 1982 that reputation was shattered.

Neither O'Kelley nor Gistaro were removed. "There were board meetings where every board member brought their own lawyer," Gistaro recalled. "There were times when O'Kelley called a board meeting, and he, I, and the board were there, and they told us we had to leave. They threw us out of the meeting because they were talking about whether to fire us. We had big-time players on the board, who had huge amounts of money and huge reputations. They could see this whole thing bringing them down."

The board included Dr. George Kozmetsky, who had been dean of the graduate of business school of the University of Texas at Austin and was later on the board of Dell Corp.; real estate developer Tom Kultznik, whom Gistaro remembered as being the largest shareholder of Aetna Life & Casualty; and Harry Bowles, who was an executive vice president at Burroughs Corp.

"Apparently they assumed that if they fired the person directly under them, someone might wonder why they, too, shouldn't be fired," Gistaro said. "So they thought they couldn't fire O'Kelley, since he was too close to them. The next guy under him, they figured he had fallen asleep, and they just demoted him. Then, under that guy, they fired five people. All of them were my friends."

Datapoint, meanwhile, was still a viable company, and could, and did, soldier on. Despite the scandal that consumed the first half of calendar 1982, its 1982 fiscal year (which ended July 31 of that year) saw revenue of $508.5 million. Somehow, revenue had risen 13.15 percent over the previous year. Many other companies would have been thrilled to report that kind of growth, but for Datapoint it was a terrible letdown from the 41 percent that it had known the previous year (and the 37 percent the year before that, and the 43 percent the year before that, etc.)[146] It remained in the black for the year, with earnings of 2.4 million, but that

---

145 After subtracting the $22.1 million from 1981 claimed revenue, the over-statement amounted to 5.2 percent, which is hardly more exciting.

146 Ironically, removing the $22.1 million revenue overstatement from fiscal 1981 would have produced a far healthier 1981-1982 revenue growth rate of 19 percent.

was down 95 percent from the previous year. There were 8,822 employees.

Datapoint carried on—but never again would Datapoint be able to conjure free money from the financial markets. Growth would no longer pay for growth (and a triumphal procession of technological advances.) Datapoint was through with Wall Street.

Unfortunately, as events would demonstrate, Wall Street was not through with Datapoint.

Chapter 16

The Raider: 1985-87

As the dust settled after the debacle, Datapoint was faced with two forms of serious damage, one obvious, the other arcane.

It would be the arcane damage that produced the most trouble—but neither would actually kill the company.

The obvious damage caused by the debacle was a loss of reputation, and the impact went beyond the stock price. Datapoint had bet on the coming of the office automation market, pushing large networks with integrated functions, including word processing and e-mail. Unfortunately, purchasing a network involved a much bigger commitment from the buyer than did the purchase of a few dedicated word processors. Buyers who had reason to question the seller's viability were slow to make that commitment.

Sales of RMS tell the tale. Only 17 installations of Datapoint's next-generation network-aware operating system had been sold inside the US by the end of 1982. Outside, the US, where Wall Street drama did not make headlines, 156 installations had been sold.

Nevertheless, Datapoint had about 20,000 users, all of whom depended on Datapoint's proprietary hardware and software to conduct business. Presumably, Datapoint could make a living servicing the needs of its existing customers, until the debacle became just a bad memory. Then, perhaps the old growth gradients could return, and it could get back into the race.

Even the Wall Street Journal's "Heard on the Street" column pointed to signs of a turnaround, noting that Datapoint's customers included 13 of the 14 largest banks in New York City, and that Datapoint had the lion's share of the remote computer market.[147]

Growth did happen—but it was not like the old days. Revenue for fiscal 1983 (which wrapped up at the end of July of that year) was $540.2 million, up 6 percent from the previous year. Datapoint got farther into the black, but not much farther, with a profit of $8.2 million.

Luxuries had to go, especially the luxury of a huge development projects involving new concepts of office automation. The ISX (the voice/data digital PBX) was canceled. There had been only one beta[148] customer anyway.

The rest of the communications management products, mostly the ACD line, were sold to another firm in May 1983.

The videoconferencing system remained in development. Meanwhile, more software was ported to RMS, and various Datapoint systems were expanded and improved—routine

---

[147] Wall Street Journal, February 15, 1983, column by R. Foster Winans. Fischer recalled that Datapoint's biggest customer in both the US and Europe were banks. However, the US banks used their Datapoint systems for office automation, while the European banks used them for transaction processing.
[148] Common in high-technology markets, a beta customer is one who agrees to test a new product, in return for discounts and other perks.

developments continued, in other words.

Datapoint's best but least-remembered achievement of the post-debacle was probably the 1984 release of the ProVista line of office automation software. Based on the network-aware RMS operating system, which allowed true multitasking, it was a non-graphics windowing system based on the DataFlash development project begun in 1982. Components included VistaWord for word processing; VistaScript for programming; VistaGuide for system control; and VistaPrint, VistaSpell, and VistaCalc for obvious purposes. It could have supported graphics-based windows similar to that used by Microsoft Windows and the Apple Macintosh, but Datapoint never came out with a graphics-based workstation for office automation. Proponents (including a Datapoint programming who soon moved to Apple) later said DataFlash should have let Datapoint come out with something like the graphics-based Apple Macintosh before Apple actually did it in 1984. Actually, that was probably not possible given that the project was started in 1982. But on the other hand the Macintosh did not become a serious office tool (with expanded storage, networking, and laser printer support) until 1987, and Datapoint could probably have beaten that schedule. Of course, such hindsight was not available at the time, and it didn't happen.

Fiscal 1984 ended with revenues of $600.2 million (the highest Datapoint would ever know) representing an increase of 11 percent. Profits had more than tripled to $28.2 million. Apparently, Datapoint had put the obvious damage behind it. Perhaps the old magic could reassert itself.

But the old magic had been based in large part on trust in the creativity of the employees, and Frassanito suddenly found that the concept of trust no longer applied. In 1984 he was marking his eighth year of running of the 25-person design firm of John Frassanito & Associates, which handled Datapoint's product development and prototyping.

One day a Datapoint staff lawyer showed up unannounced at the JF&A office to say that a Datapoint vice president had grown suspicious of the design firm's billing practices and wanted an impromptu audit of its accounting data, right then, before anyone could alter any records. Whatever he was looking for, he didn't find it, and the lawyer eventually apologized and left.

Wary of further witch-hunts, Frassanito settled his contracts with Datapoint, transferred his employees to Datapoint, and relocated to Houston. There he went back to work for the space program, development having begun for the International Space Station.

Meanwhile, Datapoint's own data was being examined by people who were finding exactly what they were looking for. The previously mentioned arcane damage suddenly began asserting itself, with disastrous results.

Arcane Damages

One of the causes of the debacle of 1982 was the arcane Wall Street concept of the price-earnings ratio. But in late 1984 Datapoint ran afoul of another arcane Wall Street concept, that of liquidation value.

Basically, even while it managed to continue conducting business successfully, the fall of its stock price in 1982 resulted in Datapoint being worth more dead than alive. In fiscal 1984 it posted "total identifiable assets" valued at $622,767,000.[149] With about 20 million shares outstanding, the liquidation value of the company was therefore a little over $31 per share. (Of course, this assumed that all corporate assets could be sold for their stated value—and events would show that getting full value was far too much to expect.) Yet, at the end of the 1984 fiscal year (at the end of July of that year) the stock was selling for only $18.75, and was falling.[150]

Depressed stock prices resulting from the dual recessions of the early 1980s had left plenty of companies in this predicament, and anyone who had the money to buy up the stock of those companies could be confident of making a profit by liquidating them.[151] Better yet, it was not necessary to buy all of the target's outstanding stock—cornering a controlling interest sufficed. And control of the company also meant control of its cash reserves. At the start of 1985, Datapoint reportedly had $107 million on hand.

And so a new breed of financier appeared, called corporate raiders, tolerated because they performed the same useful function as scavengers in the Serengeti. They had names like Carl Icahn and T. Boone Pickens. One of the lesser-known raiders was named Asher Edelman.

Age 45, usually described as brash, he was certainly handsome,[152] especially when photographed without his glasses, as he evidently preferred. His fortune was partly embodied in an art collection (which would afford him a second career, after resurgent stock prices chilled the raiding game.) His background was in finance and he had no real experience as a corporate manager—as would become painfully evident. He enjoyed skiing, and traveled around Manhattan in a Jeep driven by a chauffeur. By the end of 1984 he had already raided a couple of medium-sized firms for a profit of about $50 million, said media reports.

He had also sued his own brother as the result of a breakup of a mutually owned brokerage. He had a Brazilian wife, as well as the three children whom he didn't talk about for security reasons. At one point he was listed as having apartments in Paris and Manhattan, a Swiss residency, a Mediterranean yacht, and an Idaho driver's license.[153]

Recently he had taken over a troubled computer company called Mohawk Data Sciences. It turned out to be a debt-ridden mess saddled with obsolete inventory, whose stock price fell to $2 after he bought in at $10. Reportedly, he also had a Datapoint computer in his office.

---

149 Datapoint Corporation 1986 Annual Report, historical charts, page 37. Other sources assigned a more conservative value of $25 per share.
150 Ibid., page 46.
151 Liquidation could also take the form of reorganization, followed by the sale of subsidiaries. The raider might also simply sell his shares at a profit after the price of the stock rose because of the renewed interest in it triggered by the takeover raid itself. Or the raider might wait for the board of directors of the targeted company to pay him to go away, a practice called greenmail.
152 A widely retold tale has it that Edelman was the model for the Gordon Gekko character (played by Michael Douglas) in the 1987 Oliver Stone movie "Wall Street." Douglas won an Academy Award for his performance, which inserted into popular culture the line, "Greed, for lack of a better word, is good."
153 Background information on Edelman is derived from contemporary coverage in the New York Times, the Wall Street Journal, the San Antonio Light, and research the author conducted for several magazine articles in 1986. This included a phone interview with Edelman, in which he said that Datapoint would become a billion-dollar firm in five years, and that he loved the computer business—it was "fun." He did not respond to interview requests for this book.

So perhaps by subliminal suggestion, Datapoint (which, for all its problems, was not falling apart) eventually got his attention.

"He saw some amorphous relationship that could develop between Mohawk and Datapoint, wrapped around service," Gistaro later recalled. "He looked at Datapoint and saw that it had a customer base and a sales force and a service organization, and was sure that he could do something with it."

In November 1984, the price of Datapoint stock began to rise, reaching $14, indicating that somebody, somewhere, was buying up large blocks.

Then, on December 9, the truth came out—Edelman filed required SEC papers indicating that he now owned more than 5 percent of Datapoint's stock, and was interested in pursuing a liquidation, sale, or merger, etc. The stock price gradually rose above $20 as Edelman soon acquired 10.8 percent of the outstanding stock.

Datapoint's management was aghast—suddenly, the company appeared doomed. They rejected Edelman's initial takeover bid and set out to find a "white knight" who would buy the company and not liquidate it. Supposedly, they approached as many as 100 different companies (although that may be counting all talks held since the debacle in 1982) but experienced no luck. Suitors criticized it for its late entry into the 32-bit microprocessor field (which happened in November 1984) and (as Gistaro noted) for its tardiness in making the ARCNET compatible with the PC. The only part of the company that anyone expressed interest in was the customer service network.

Stock analysts were also aghast, since Edelman's move did not entirely make sense to them. The raiding game worked fine with companies whose assets had established value, like factories, stores, ships, mines, timber, or real estate. But the assets of a computer company are its technology (which is derived from its skilled employees), its skilled employees (who can walk away), and its customers (who can also walk away).

Customers did, indeed, start walking away, on the assumption that Datapoint had no future except liquidation. The firm lost almost $16 million for the quarter the ended in January 1985, and laid off 659 people.

Edelman launched a proxy battle for control of the company, mailing out ballots to the other stockholders to elect a new board of directors. No one expected him to lose, so in mid-March O'Kelley and Gistaro went to New York to negotiate their surrender directly with Edelman.

"O'Kelley did not want to go through with a proxy fight," Gistaro recalled. "I was acting as a messenger boy between our board, in a hotel, and Edelman's office. Edelman would say, tell the board this, and I would go back and tell them. Finally we got all the terms ironed out that way, via feet mail. There were no e-mails or phone calls, it was just me. It was all done in the course of one day."

They reached a compromise in which the incumbent board members would not simply be turned out. Instead, the number of seats on the board was expanded and equally divided between Edelman's associates and the previous members. O'Kelley himself would have to step down.

"O'Kelley had to leave, but he had already said that he wanted to," Gistaro recalled. "Edelman knew full well that my board members were not going to stay. I pleaded with them to stay because I did not want to be the only guy left standing, but every one of them resigned shortly thereafter, although not immediately. They were probably breathing a sigh of relief that they never incurred any liability through all this mess." As for Edelman's board members, "They knew less about the computer industry than he did," Gistaro added.

O'Kelley, stunned, could not find his car when he got back to the San Antonio airport.

"He was devastated," recalled Gistaro. "He had come to this city, where he was not initially socially accepted, and had run its largest industrial employer, by far. The old guard in San Antonio—few of them made their money by themselves—had treated him like some bumpkin who had wandered into town."[154]

O'Kelley would retire to Indialantic, Florida (just south of Cape Canaveral, where his high technology career began) and resume his hobbies of genealogy and watercolors. He died there on December 8, 2000.

The Alternative Ending

At the time of his takeover, Edelman had apparently planned to break Datapoint into three pieces—service, manufacturing, and international sales—and sell them separately, hopefully for a profit. Repeating the recent experience of the O'Kelley administration, he apparently spent two weeks finding no interest for the manufacturing and the international marketing divisions. As for the service operation, it was Datapoint's most profitable division.[155]

With this experience behind him, Edelman made his first visit to Datapoint on March 28. He toured the facilities, met with various officials, and saw products that were under development.

"Edelman loved the videoconferencing system," Gistaro recalled.[156] "But what impressed him when he first visited Datapoint was the kind of people he talked to. I don't remember exactly who they were, but we had some very smart people at Datapoint. The kind of people he normally dealt with were not smart in the way these people were smart."

At some point he also met the mayor[157] of San Antonio, who made a pitch about how important the company was to the city.

The next day he held a press conference—and announced that he was not going to

---

154 Gistaro recalled that his own application for membership in a local country club was answered with a one-line acknowledgement, and then 17 years of silence. Finally, a sponsor got him accepted.

155 Datamation, May 1, 1985.

156 This was the same system that David Monroe began tinkering with after joining Mnemonics in 1974, with the help of Frassanito, as noted in Chapter 11. After Mnemonics collapsed he returned to Datapoint with a hiring contract that recognized his ownership of the technology. Before he left during the debacle in 1982, he had developed it into a fully functional system that shared Datapoint's network. What happened after that is covered in Appendix A.

157 San Antonio's mayor at the time was Henry Cisneros, the first Hispanic mayor of a major US city in the 20th Century. Later, during the Clinton Administration, he was US Secretary of Housing and Urban Development for four years until a scandal involving his personal life led to his removal.

liquidate Datapoint. "I don't melt steel or people," he was quoted as saying. He said he was impressed with the upcoming products he had previewed, and with the people he had met.

The stock fell $3. Wall Street analysts were aghast—Edelman's rational move was to liquidate, and he wasn't doing it. As with scavengers in the jungle, there was no sense being sentimental.

Datapoint insiders were also aghast, as word spread that what had changed his mind was the previously mentioned network-based full-motion, full-screen color videoconferencing system.[158] There was no secret about the system, which was treated as an executive toy.

Meanwhile, Datapoint was still in play. Edelman soon announced that Datapoint would spin out a new, separate company from its service division, under the name Intelogic Trace.[159] Datapoint stockholders would get a share of Intelogic Trace for every share they owned of Datapoint.

A new chief operating officer for Datapoint was hired in May. Four weeks later he was fired, apparently because he opposed the spin-off of the service division. He sent a parting letter to the board of directors questioning whether Datapoint would remain viable after the spin-off. Edelman proceeded with the plan anyway.[160]

The damage caused by all the turmoil was evident after the end of the fiscal year on July 31. Revenue was $520.2 million, down 13 percent from the previous year. The magic of eternal growth was gone. Worse yet, the company lost $48 million. It was the first annual loss since 1972.

As for the videoconferencing system, it was turned into a product by the end of the fiscal year, under the name MINX (Multimedia Information Network eXchange.) It was the first desktop videoconferencing system ever marketed. Units cost $8,800 to $11,000, depending on quantity. The government turned out to the biggest buyer, as it was already a big user of networks. In the next two years Datapoint sold about a hundred systems. But (as discussed in Appendix A) in the process of marketing MINX, Datapoint unknowingly managed to sabotage a patent infringement case it would be involved in a decade later. Had it not been for that blunder, it may have made more off the lawsuit than it ever made from selling MINX.

The Edelman Administration

For the rest of its existence, Datapoint would be tied to Asher Edelman.

"I guess I met Edelman for the first time face-to-face when he first came down to visit

---

158 After the videoconferencing system was put on the market, Edelman would deny that it was what changed his mind about liquidating Datapoint, telling an interviewer that there was another upcoming "breakthrough product" in the wings. (Express-News, July 31, 1985.) However, there is no indication that there was any such additional product. Fischer recalled that Doris Bencsik, then a vice president, claimed to have convinced Edelman that liquidating was not in his best financial interest.
159 Reportedly, the name was chosen at the last minute in order to claim the IT stock symbol, which had become available. Two-letter symbols were considered prestigious, and IT also stands for information technology.
160 Aside from stripping Datapoint of its service revenue, the plan required that Intelogic Trace guarantee that it would continue servicing Datapoint hardware, but at inflated rates, Fischer recalled. Therefore, Datapoint had an incentive to minimize its use of Intelogic Trace, while Intelogic Trace had to remain geared up to offer full support to Datapoint.

Datapoint," Gistaro recalled. "He said, 'Okay Ed, you're the CEO,' and paid me my bonus for the prior year, which I had earned. Everything was reasonably smooth between us. But trying to run a technology company when the guy calling the shots is a raider who is probably going to break up the assets is pretty damn tough. How can you go into a Citibank and say, 'We'll be around to support you,' and they say, 'Aren't you owned by Asher Edelman?'

"I used to tell people that the biggest problem with dealing with Edelman is that if it's Monday he thinks long-term is Thursday," Gistaro said. "I think he was interested in operational results (opposed to pure financial manipulation.) In fact, I think he got sort of enamored with the whole thing. It was the first time he had ever been involved with anything that had any substance. He had never been involved with a company that was actually running, and had interesting people, and interesting products, and revenue, and all that jazz. He got beyond the numbers, but he was still a numbers guy.

"This was not a subtle man. He was outrageous. He would not talk about the things he was really an expert at, because he did not want to give away anything. And when he did talk about anything else he was full of nonsense. He'd talk about products and markets, but he wouldn't know what he was talking about, everything was a bit large with him, everything was massive in scope. He exaggerated a lot about anything, about how good someone was, about how bad someone was, how good a product was, how bad a product was, how good a company was, how bad the company was. It was all over-blown. A psychiatrist would leave mumbling to himself.

"The things he would do and say in board meetings were just outrageous," Gistaro continued. "He would say, 'We can tell the shareholders this, they can take a joke.' I can't remember specifics but I remember being flummoxed. I came from a board of directors that were leading lights of the industry, to a board of directors who had no experience in anything, one ran a lumber company, another was a used-car salesman.

"I once asked Edelman if he could ski, and he said he was a great skier. I asked him how he got to be a great skier, and he said he had hired an Austrian ski instructor for a year. I went to dinner once at his apartment. He had the whole floor of the building. The dining room had the front half of a Chevrolet coming through the wall. I was looking at a large picture in his art collection and didn't really like it but noticed the elaborate carving of the wooden frame. I looked at it closely and saw that it was entirely representations of penises."

At the end of 1985, Forbes ran an article noting that a number of corporate raiders were stuck running the companies they had acquired, and most were floundering. As for Edelman, he bought in for an average of $20 per share, but by the time the article was written the combined stock prices of Datapoint ($5.50) and Intelogic Trace ($8) was only $13.50.

In other words, if he sold out he would suffer massive losses. His only hope for getting clear was to keep the company in the black and restore a respectable level of growth, until the stock price rose again.

As 1985 stretched into 1986, Edelman launched an effort to buy out the rest of Datapoint's outstanding stock at $6 per share, but dropped the idea when financing proved elusive. When there were questions about his use of Datapoint and Intelogic Trace pension funds to buy stock in his latest takeover target, Edelman said it was normal investment activity.[161]

At the end of the 1986 fiscal year, in July, Datapoint posted revenue of $325.23 million. That was nearly $200 million less than the previous year, but much of the difference was accounted for by the spin-off of Intelogic Trace, which took with it all of Datapoint's service revenue. More importantly, Datapoint lost $8.56 million.

Gistaro Departs

One thing that Datapoint had never done was pay dividends. Investors profited from owning Datapoint stock when the price went up and they were able to sell at a profit. By 1986 that was no longer a regular event, and investors were better off keeping their money in the bank.

So in the latter part of 1986, Edelman approached Gistaro with the idea of creating a new grade of Datapoint stock: preferred stock paying a 13 percent annual return. Preferred stock does not give the owner voting privileges like common stock, but its dividends (or liquidation claims) must be paid before any other financial obligations.

"Edelman was having trouble with his investment group," Gistaro recalled. "They weren't making money. They thought he would go in, split up Datapoint's assets, sell everything off, write them a big check, and go on. But he didn't do it. He had to make them happy so he decided he was going to let them convert their common shares into preferred shares.

"I said, 'Why would we do that? The last thing this company needs right now is another financial obligation on top of all these other problems.' He said that if we can give them a little coupon they would cool off and not be so unhappy.

"I said this would be a really hard sell to the investment community since they will say that we are bailing on our common stock. He said, 'Tell them that I'm not doing it, and I'm the biggest shareholder.' So I went out on the stump like an idiot and made speech after speech to the local press, and investor groups, and customers, and anyone I could get to listen. I said, 'Edelman is not doing it himself. If the largest share holder still has confidence in the common stock, you can have confidence in your common stock.' They bought it, to a certain extent.

"So we filed the required form with the SEC saying this was going to happen. And it said that Edelman was not going to do it, that Edelman was not going to convert his common stock. So we were happily going on our way and the final deal happens, the final filing comes out—and my chief legal officer runs into my office saying, "Ed, look at this, Edelman converted all his common stock.'"

Gistaro called Edelman's personal attorney in New York. In the process, Gistaro later recalled, he found that Edelman's lawyer was not covered by his own law firm's liability insurance policy—he was excluded specifically because he was Edelman's lawyer.

Gistaro told Edelman's lawyer what he thought, and the lawyer passed it on to Edelman, who came down to San Antonio soon thereafter and took Gistaro out to lunch.

---

161 Gistaro said he lent Edelman some of his own funds at one point, and made a profit.

"He said, 'Ed, I think it might be time for us to part ways.' I said, 'Yeah, you're probably right, Asher.' And so we did. There was no yelling at all. I would have had to leave soon anyway because I was really swimming upstream and not being a salmon I couldn't pull it off."

Gistaro's departure was announced in the Wall Street Journal on January 16, 1987. Doris Bencsik, who was executive vice president and chief operating operator, was made acting CEO.

Subsequently, Edelman would claim that it was all a misunderstanding caused by Gistaro's legal counsel, who (Edelman said) drew up the SEC filings without showing them to Edelman. That man then sued Edelman for slander and collected $3.6 million, Gistaro recalled. And while Edelman tried to get out of paying Gistaro the three years' severance pay that he was due, Gistaro's employment contract, written by a San Antonio attorney, proved iron-clad, Gistaro added.

After Gistaro's departure, Datapoint would last for more than a decade. There would be brief periods of stability, followed by periods of steep declines. But the forces that would destroy it were already in full play.

Chapter 17

Converging Dangers

The debacle and subsequent loss of reputation, stock price collapse, and raider takeover were all damaging to Datapoint. But that series of events is not what destroyed the company. Corporations have muddled through worse—consider Ford and the Edsel, or Coca-Cola and New Coke, or Apple during the absence of Steve Jobs from 1985 to 1997. Had Datapoint continued to grow like it did during the 1970s the debacle would have become a troubling but fading memory. Edelman's involvement would have been even more forgettable, as he would have surely sold his stake for a profit after the stock price recovered.

But as the firm gradually emerged from these crises in the mid to late 1980s it faced two trends that would converge to destroy it. The first was short-term, and some mitigation proved possible. The second was long-term, and it is hard to imagine that Datapoint could have done anything about it.

The short-term problem was that, by the mid 1980s, Datapoint's core product line, its processors, was no longer competitive with those of other computer vendors.

The long-term problem was that the computer market would eventually drown in an ocean of inexpensive PCs, and Datapoint's hard-won position in that market would become irrelevant.

Let's deal with the short-term problem, and then the long-term one.

Short-Term: the Drought

Again, Datapoint's short-term problem in the mid-1980s that was its processors were no longer competitive with the rest of the computer industry. Hired by Datapoint in 1985 as its principal product architect, Michael Fischer blamed the situation on what he called Datapoint's "processor drought." This section is based on his recollections, and archives.

The processor was the core of a computer vendor's product line. The vendor's operating system was designed to run on it, and various peripherals like terminals and printers were optimized to run with it. Application software, meanwhile, had to be designed to run on the operating system, which depended on the processor. Obviously, a vendor's success or failure hinged on the power and capabilities of its processors. So let's summarize CTC/Datapoint's offerings, ignoring the Z80-based 1500, the MINX videoconferencing system, and various peripherals. (A more technical comparison, with data on competing processor chips and PCs, is in Appendix D.)

Datapoint 2200, 1970
This was the original one-bit machine whose instruction set was carried forward by subsequent Datapoint processors, as well as, in increasingly modified form, the derivative x86 processor dynasty that dominates the digital landscape today. It had a half-height screen and supported 8K of memory.

Datapoint 2200 Version II, 1972
Using the same screen and enclosure as the original 2200, this version was a true eight-bit machine, running at 2 megahertz, supporting 16K of RAM.

Datapoint 5500, 1975
Still using the enclosure and half-height screen of the 2200, the 5500 ran at 4 megahertz, supporting 64K of RAM. It was commonly referred to as a 16-bit machine, moving Datapoint into the computer market. However, while some 16-bit instructions had been added to the processor's instruction set, it still largely relied on 8-bit operations. The 5500, incidentally, came out three years before Intel's 16-bit version of the same dynasty, the 8086.

Datapoint 6600, 1977
Still retaining the enclosure and screen of the 2200, the 6600 was a 16-bit machine running at 6.67 megahertz that used segmented memory to support up to 256K of RAM. Each instruction required fewer clock cycles to execute than the 5500, and on the whole the 6600 compared favorably to the Intel 80286, which came out five years later, in 1982. Indeed, the main difference between the 6600 of 1977 and the 80286-based IBM PC/AT that came out seven years later in 1984 is that the latter included a crude graphics display.

Datapoint 1800, 1978
This desktop unit with a full-height screen and detachable keyboard was intended for use in remote offices for data entry, sending files to the main office via a telephone modem. It had a modified 5500 processor supporting 64K of RAM, running at 2.5 megahertz, using 8-inch diskettes for storage.

Datapoint 3800, 1979
Outwardly similar to the 1800, the 3800 was intended for office use (especially for Datapoint's IEOS office automation software that came out the same year) and required an attachment to an ARC network for file storage. It used a modified 5500 processor running at 2.5 megahertz, supporting up to 128K of RAM.

Datapoint 8800, 1980
This was a rack-mounted heavily customizable machine in a stylish enclosure about the size of a three-drawer filing cabinet. It could support a megabyte of RAM and a gigabyte of disk space, and had a high-speed disk interface, and supported multiple ARC interfaces and terminal interfaces. (It did not have a built-in screen or keyboard, depending on an attached terminal for control.) Unfortunately, it had the same 16-bit performance as the 6600 and ran at only 5 megahertz. Fischer felt that any system that came out in 1980 should (and could) have had twice the power that the 8800 offered. Reportedly, many buyers were disappointed. Beyond that, its operating system initially was the newly introduced network-friendly RMS, freezing out the bulk of Datapoint's customers, who were still using Datapoint's previous operating system (DOS).

Datapoint 8600, 1981
This 16-bit desktop unit, running at 4 megahertz and supporting 512K of memory, had an

integral full-height screen and was intended for office automation. It had no local storage, instead using the ARC network for disk access. It was intended to use a single-chip version of the 6600 processor called the Redwood chip, but the effort was abandoned when the selected chip fabricator (primarily a memory maker for video games) failed at the task. So a board-level processor had to replace a chip-level processor, as had happened with the 2200. This delayed introduction by a year. An 8 megahertz version came out in 1985 to compete with the IBM PC/AT (which initially ran at 6 megahertz) after it was noticed that the original design of the 8600 actually supported 8 megahertz.

Datapoint 8400, 1984

This was a version of the 8600 that used a 6 megahertz Intel 80286 microprocessor paired with circuitry called the TLX Engine to execute the Datapoint instruction set. This adoption of an Intel chip offered economic advantages to Datapoint but in this case did not increase performance significantly. It ran Datapoint's RMS-based PRO-VISTA office automation system that came out that year, described in Chapter 16.

StarShip I, 1985

Otherwise known as the 8850, this processor was, basically, an upgraded 8800 supporting DOS, introduced in an effort to match the machines being marketed by Performance Technology, a firm described in the next section. DOS could just as easily been included with the 8800 line in 1980, and a similar upgrade would have been expected in 1982.

Datapoint 7600, 1986

This was a repackaged, reduced-cost version of the 7400.

StarShip II and III, 1986 to 1990

Otherwise known as the 7000-series, the later StarShip line (with no real connection to the first StarShip line) was intended to replace the obsolete Datapoint processor line over a multi-year period, with development starting in 1985. It fulfilled its aims, despite the loss of half its development staff and budget along the way. The 7000 line used between one and four of the latest Intel processor chips to achieve a wide range of power among machines that could all nevertheless run the same software. They were designed to support faster Intel processors as they came out, and managed to do this for three processor chip generations. RMS/XA, the version of RMS that survived Datapoint, also came out during this period.

Clearly, leading into the early 1980s Datapoint's dynasty of flagship processors had been the 2200, 5500, 6600, and the 8800. But their clock speeds had not increased significantly since they reached 4 megahertz in 1975. Beyond that, while some existing processors were repackaged, the list shows that there were no significant new processors between the 8800 in 1980 and the Starship II line in 1986. Six years is a long time in the computer industry. The IBM PC went through three generations during that period (the PC, the PC/XT, and the PC/AT.) During that time, Datapoint's revenue came mostly from machines that were engineered in the 1970s. During the 1970s, Datapoint was typically several years ahead of the rest of the industry (as much as ten, if you count office networking) so it had some cushion. But there was little improvement in the 8800 over the 6600, and after the 8800 there seemed to be little to show for Datapoint's development efforts—which nevertheless continued at their previous pace.

So why the drought? Sources point at several factors, and the final answer is doubtless a combination of the all of them.

First, the phenomenal revenue growth that Datapoint experienced in the 1970s also required phenomenal growth in the number of employees, and managers. While this influx was underway there had evidently not been sufficient planning for management growth and decision making (assuming such planning was even possible.) Inter-departmental feuding was common, and the bureaucracy had bloomed to the point where 17 people had to sign off on an ordinary press release before it could be sent out. Seeing any project through to completion required a triumph of personal stubbornness. Basically, growth had brought its own form of management chaos.

It was as if the groupthink metaphors from Kafka, which Roche had discussed with Frassanito when they first met in 1968, had returned to haunt Roche's creation.

Of course, the stock market debacle and the takeover crisis did not help. Management's attention was diverted to topics that had nothing to do with the company's product line.

But keep in mind that the drought was not necessarily as painful at the time as it seems in hindsight. As discussed in the next chapter, keeping ahead of the curve, technologically, is not necessarily the key to success in the computer industry, or any industry. It's probable that Datapoint's offerings were still well ahead of what most of its customers wanted or could use, and the ARC system let them apply more processors to a task when the previous installation proved inadequate. If still more power was needed, the plan in 1981 was to offer even more processor connectivity, though the ISX digital PBX. Computational power would be out there, somewhere, on the network. Users with bogged-down applications no longer had to wait for Moore's Law to rescue them.

As for the ISX project, doubtless it seemed like a good idea at the time, and was years ahead of the market. Very few potential customers had heard of a local area network in 1981, but they all had office phone networks. Meanwhile, Datapoint already had some presence in the office telecommunications market thanks to its ACD (automatic call distributor) product line. By developing a digital PBX that could connect terminals to an ARC network, Datapoint could merge the new world of office data communications with the established world of office voice communications. But Datapoint found that corporate managers of those two worlds rarely talked to each other and certainly did not pool their purchasing budgets. In the end Datapoint diverted large amounts of development resources away from products with proven revenue potential, to a project that never made Datapoint a cent. Reportedly, for several critical years in the early 1980s more R&D funds were spent on the ISX then on Datapoint's core line of processors.

The launch of the ISX project, meanwhile, was a symptom of a lack of focus throughout the company, as a cohort of newly arrived managers struggled with the question of whether Datapoint was now supposed to be a computer company, an office automation company, a networking company, a telephone system company, or the next IBM.

But thanks to the long-term fatal danger posed by the newly arrived microcomputers, the answer didn't matter.

Lateral Move

As for what Datapoint could have been doing during the drought, there's the example of Performance Technology, a firm started by former Datapoint employees in 1985. Basically, one of the cost-cutting measures taken after the Edelman takeover that year was to eliminate the company's advanced products development group, led by Jonathan Schmidt. Given that every previous success that Datapoint could claim stemmed directly from its product development efforts, this move did not bode well for the future—and what happened next was actually embarrassing.

Schmidt and a number of other staff members simply made a lateral move to their own startup and began making computers based on carefully selected, high-speed off-the-shelf components, including the latest microprocessors. Using specially written software that avoided violating any Datapoint copyrights, these machines could run programs written for Datapoint machines—but ran them much faster than any Datapoint hardware could. Plus the machines were physically smaller.

In other words, they were doing everything Datapoint should have been doing.

Performance Technology began operations in August 1985, and its first sales were in November. It was profitable after four months and grossed $36 million its first year.[162]

Schmidt recalled visiting the offices of the French carmaker Citroen in Paris. He was shown a room full of Datapoint servers that handled the overnight financial reconciliation function. Unfortunately, the assembled machines were taking more than 24 hours to complete the task, so the accountants never caught up.

Schmidt brought in a Performance Technology demo machine that fit on a shelf. He installed it and transferred the reconciliation software and database to it. It fit. He began running the reconciliation process. After some small talk with the Citroen computer manager, he was about to leave, when the machine made a different noise. Schmidt checked it and saw that it had already finished the process—taking a few minutes, as opposed to more than 24 hours.

Another example from press coverage involved an Australian travel agency that took 17 hours to do overnight financial processing using Datapoint servers. With a Performance Technology machine they could do it in 45 minutes, and were able to lay off their night shift computer staff.[163]

Initially seeing the newcomer as a threat, Datapoint finally agreed to resell Performance Technology products in November 1986. That, and the introduction of the Starship II line, helped restore Datapoint's position in the market—a market about to be destroyed by

---

162 San Antonio Light, March 3, 1987.
163 Performance Technology was acquired by Bay Networks in 1996, and Bay Networks was acquired by Canadian telecommunications conglomerate Nortel in 1998. Nortel was forced to restructure after the dot-com crash in 2001, folding the Performance Technology division. The employees reformed under the name Pertech, which had been the Internet domain name of Performance Technology. They continued operating in San Antonio, selling software for broadband Internet suppliers. As for Nortel, it was plagued by accounting scandals and began liquidating in 2009.

the previously mentioned long-term threat. Ironically, the threat owed its existence to Datapoint's accomplishments in the early 1970s.

Long Term: the PC Tsunami

Microprocessors had, as described, began appearing in 1972, first in specialized devices such as blood analyzers. Due to its clumsy multi-chip interface requirements, the original Datapoint-derived 8008 chip was not really suitable for an inexpensive computer. But is successor, the 8080, solved many of these problems when it came out in April 1974. Small computers based on it (called microcomputers to distinguish them from the mainframe and minicomputers that preceded them) began appearing in 1975. They were typically hobby kits, costing a few hundred dollars. The absence of significant software was not an issue since writing software was part of the challenge for the hobbyist.

The most famous kit was the Altair 8800, announced in the January 1975 issue of Popular Electronics. For $397, the buyer got an 8080 CPU with a paltry 256 individual bytes of memory. With that amount of RAM—more could be added—the machine could run a program that could change the pattern of the indicator lights on the front panel.

But despite their limitations, about 50,000 microcomputers were sold worldwide that year, hinting that the public harbored a voracious, unfilled hunger for computing power.

By contrast, in 1974, the total number of minicomputers in use worldwide was estimated to be 150,000. At the end of October 1974, Datapoint had proudly announced that it had sold, to then, 3,971 Datapoint 2200 desktop computers to a total of 409 customers.[164]

(A leading vendor of kit microcomputers was, oddly enough, a San Antonio firm that had been operating half a block down and across the street from CTC's first headquarters on West Rhapsody. Called Southwest Technical Products Co. or SWTPC, it got started in audio systems but had expanded to kit computers by late 1975. By 1991 it, too, had been driven out of the market by the PC tidal wave.)[165]

In 1977, mass-produced, assembled and tested personal computers that could run useful software began to flood the market, such as the TRS-80, the Commodore PET, and the Apple II. Someone who wanted to own a personal microcomputer no longer had to assemble a kit with a soldering iron. However, models from different vendors could not run each other's software. In fact, this could be true for different models from the same vendor. But these machines whetted the public's appetite.

Worldwide microcomputer sales reached 180,000 in 1977, rising to 330,000 in 1978, 620,000 in 1979, 1,205,000 in 1980, and 1,850,000 in 1981.

Late 1979 saw the arrival of VisiCalc, the first desktop spreadsheet, running on the Apple II. This gave businesses a legitimate reason to buy them. Of the 130,000 Apple II units sold before September 1980, an estimated 25,000 were sold primarily to run VisiCalc. (It also

---

164 San Antonio Express News, November 13, 1977)
165 See http://www.swtpc.com/mholley/History/SWTPC_History.htm, accessed March 25, 2009. The author was one of their customers in the late 1970s, building a machine that was later given to the Smithsonian. Ironically, SWTPC used Motorola CPUs, rather than the Intel chips whose ancestor owed its existence to what was going on across the street.

triggered the sale of a lot of RAM chips, as VisiCalc consumed a then impressive 25,000 bytes of memory.)

The response of Datapoint and its competitors was to dismiss these machines as toys.[166] After all, the new personal computers made no claim to having the kind of engineered quality that the full-service computer vendors offered for both their hardware and software. If something went wrong, a trained technician was not going to show up and fix it.

"We are businessmen selling business systems to the business community and we do not make things that go under Christmas trees," Frassanito remembered being told.

But those microcomputers under the Christmas tree did spawn a cottage industry of small startups producing add-on components and application software, offering steadily improved word processors and spreadsheets.

The difference between the home and the business computer market was highlighted by a misunderstanding that supposedly caused a former Datapoint programmer to nearly wreck Radio Shack's efforts to offer a disk operating system for the TRS-80, so it could use disk drives instead of cassette tapes for mass storage.

David Monroe was his office mate at the time, and recalled the programmer loading large amounts of Datapoint equipment, software, and documentation into a vehicle and driving away, after explaining that Jonathan Schmidt had authorized its removal for off-site work. A couple of hours later he approached Schmidt with a requisition to replace the removed items, and found that there had been no such authorization. Through a later lawsuit they were able to recover the hardware, but the judge ruled that the intellectual property had no legal protection, since Datapoint itself had not adequately protected it.

The programmer moved north to the Dallas area and was paid to write TRSDOS 1.0 in about 1978. He apparently assumed that Radio Shack had the same kind of quality assurance and testing functions that he had worked with at Datapoint. It didn't, and when he submitted his first effort the company blithely turned around and sent it out to thousands of customers, where it promptly crashed. The software's bugs were eventually fixed, but there was no way to repair its reputation.[167]

The TRS-80, incidentally, was based on the Faggin's Z80 microprocessor, and was therefore another descendent of the Datapoint 2200. Radio Shack was able to sell about a quarter million of them between 1977 and 1981, with a retail price starting at $600.[168]

David Monroe, among others, argued that Datapoint ought to port its software to the new machines. However, they were daunted by the potential expense, as it was written in assembler rather than a higher-level language.

Monroe recalled that he developed a ROM, costing about $2, with code to perform Z80 hardware instructions not supported by a Datapoint processor. This would circumvent the need to translate Datapoint's software. He remembered demonstrating to Gistaro that

---

166 This was explained at length in the Datapoint sales training class attended by the author in 1980.
167 See "Priming the Pump," by David and Theresa Welsh, published in 2007 by The Seeker Books, Ferndale, MI. The book covers the history of the TRS-80.
168 See http://en.wikipedia.org/wiki/TRS-80, accessed December 14, 2009.

this ROM would let them produce a machine that could sell for less than $3,000. Gistaro responded that they were not making computers "for hippies to use in their garages," Monroe recalled. Nothing was done.

On the other hand, the Z80 was used as the basis of the previously described Datapoint 1500, with a subset Datapoint software that was translated for the purpose. Intended for satellite offices, it initially lacked an ARC interface. A version of CP/M (an industry-standard operating system for the first-generation microcomputers) was also offered. In 1980 a 1500 with 60K of RAM and two floppies cost $7,825.

In four years Datapoint would sell about 6,000 units.[169]

Against the Tide

But in 1980, unheralded events began to unfold that would make all previous computer sales numbers seem like a flyspeck, and seal Datapoint's doom. In that year Intel launched Operation Crush. The idea was to eliminate the threat that the new hybrid 16/32-bit Motorola 68000 microprocessor posed to the Intel 8086, the 16-bit version of the x86 family that was released in 1978. By this time, Intel had figured out how to market microprocessors. The idea was to convince companies that made computers and other high-tech devices to use Intel's processor in their new designs, by smothering them with attention, samples, and engineering help. Successful efforts were called design wins. Operation Crush was launched with the intention of securing 2,000 design wins. In the end they got closer to 2,500, and in a few years the Intel line was out-selling the Motorola line by nine to one.

One of those design wins was IBM, with its new Entry Systems Division. Not only had Big Blue decided to enter the microcomputer market, but it had decided to break with the past while doing so. Traditionally, IBM relied on vertical integration, meaning that it made all the parts that went into an IBM product. The time required to tool-up meant that getting a new product to market took three years. IBM's managers realized that the pace of progress in the microcomputer field made a three-year delay unacceptable. So they traded profit margin for speed by designing their microcomputer largely around off-the-shelf parts from third parties. They chose the Intel microprocessor line because of its popularity, and because Intel's planned upgrades would remain backward compatible with the current generation. That was not the case with Motorola, Intel's main competitor at the time. A 32-person firm in Seattle that sold microcomputer software, called Microsoft, was contracted to do the operating system. IBM also chose to sell the new machine through retail stores (including its own) as well as through its direct sales force.

There is no record that Datapoint even reacted when, on August 12, 1981 (i.e., 6 months before the debacle) IBM unveiled its new machine, grandly named the IBM Personal Computer, or PC. A unit with 64K of RAM and a floppy disk cost $2,880. But instead of using the 16-bit 8086, it used a derivative that came out in 1979 called the 8088. The 8088 chip

---

169 Fischer recalled that a version of the 1500 called the 1569 with an ARC interface but no disk, and with an intended price of $1,500, was developed in 1982 and 1983 but never marketed, engendering considerable bitterness among those who had worked on it. He felt it could have sold 6,000 per quarter, if not per month. The 1569 could also have been the first computer to use a motherboard chip set, where all the main functions (other than memory) are concentrated in a handful of standard chips. The industry caught up in 1986.

did, indeed, have a data path inside the chip that was 16 bits wide and performed 16-bit operations (as did the processors that Datapoint was making at the time.) But, as was previously mentioned, it used an 8-bit data bus (the connection used to control components outside the chip) throughout the rest of the computer. This allowed the use of cheaper, plentiful 8-bit components, but also slowed down the machine's overall operation. (It made them slower than Datapoint's machines, which had processor speeds about equal to that of the original PC, but had 16-bit buses.)

Consequently, the IBM PC was, technologically, not very interesting—except that we're talking about IBM, whose strategy was to sell reliable products with good profit margins, incorporating technology that was just good enough to dominate a market when IBM's marketing clout was brought into play. Good enough technology would take a back seat to best-of-breed marketing, in other words. Moreover, IBM's unprecedented break with its usual strategies (such as the use of the retail sales channel, and the use of third-party components, which themselves could accept third-party add-ons) hinted that this was a product that could change the industry.

IBM launched an unprecedented national advertising campaign, using a Charlie Chapin look-alike. Mere days after the PC's announcement, IBM decided to quadruple production, as the retailers could not keep the machines in stock. By December, IBM had shipped 13,000 PCs. In the next two years it would sell about half a million.[170]

Equally significant was the fact that business buyers saw IBM's involvement as an endorsement of the microcomputer concept, and ceased seeing them as toys. If they had qualms about reliability, they bought spares. After all, the PCs were often cheaper than the desks they sat on. The PC's operating system, PC-DOS, was seized on as a much-needed standard in a market swarming with systems that were incompatible with each other. Immediately, other vendors sought ways to "clone" the IBM PC in ways that did not violate any intellectual property. The first truly successful clone-maker, Compaq Computer Corp., was launched in Houston in 1982, and had sales of $110 million in its first year. Clones ran a generic version of PC-DOS that came straight from Microsoft, called MS-DOS. More importantly, hundreds of vendors rushed to offer hardware and software that was "PC compatible," creating an ecosystem that dwarfed the cottage industries that had served the various eight-bit consumer-oriented machines that had come before.[171]

Also just before the debacle, David Monroe became involved in a lengthy effort, with Frassanito, to plot the direction of Datapoint product development a couple of generations ahead, using the wall of an empty office. He recalled that he was trying to promote the kind of information transferability now taken for granted. For instance, he wanted to work toward a fax on the ARC, so that a word processing document could be sent directly to a fax machine, or an incoming fax could be sent directly to a laser printer or to a computer. Having built an enterprise-class laser printer, he foresaw a personal laser printer. Frassanito recalled Gistaro coming by, examining the wall, and leaving without a word. Monroe recalled making presentations based on it, for O'Kelley and other officers, using Datapoint's new Color

---

170 PC shipment numbers from "Fire in the Valley," Paul Freiberger and Michael Swaine, 1984, McGraw-Hill Inc., page 279.
171 PC history information is from "Computer: a history of the information machine," Martin Campbell-Kelly and William Aspray, 1996, Basic Books, New York.

Business Graphics system. Thanks to the PC tsunami, any response he got soon became moot.

While the PC tsunami took shape, the Apple Macintosh came out in early 1984 and quickly became the first commercially successful personal computer based on a graphical user interface. Hypertext inventor Ted Nelson, then employed at Datapoint, recalled being in a store that sold Macintoshes near Datapoint's headquarters when he saw Datapoint CEO Harold E. O'Kelley walk in, alone.

"In the way he was walking I saw fear, determination, and bravado," Nelson recalled. "He knew he had to see this thing but did not know what it would be—he was way behind the curve."

But whatever else he saw there, O'Kelley would also have confirmed that, for the Macintosh, as with the PC and its clones, service meant taking the machine back to the store and hoping that someone there could fix it. Datapoint and its competitors could still confidently point to the advantages they offered over the microcomputers in terms of reliability and service. A business that had computerized its operations needed reliable hardware, not something designed for the home market. Entrenched customers continued buying premium computers, and so the premium prices charged by Datapoint and its competitors appeared justified—for the moment.

Frassanito recalled showing one of the first Macintoshes to Vic Poor, who shrugged. In five years, he predicted, no one would remember Apple.

In 1984 Datapoint (following the half-hearted steps of other computer companies) began selling a PC-like desktop machine under its own name, a rebranded NGEN from Convergent Technologies. It was labeled the Datapoint 1600. As with the other computer firms, analysts complained that Datapoint's response to the PC was over-priced and not fully MS-DOS compatible.

When IBM came out with the third-generation PC that year, the PC/AT, Datapoint issued a press release noting that its own PC was cheaper than the IBM PC/AT and therefore offered superior price-performance. It was embarrassing—apparently no one in authority realized that the new PC/AT was based on the 80286, making it significantly more powerful than any first-generation 8088-based PC or clone (such as Datapoint's.) The PC/AT actually had a clear price-performance advantage.

In early 1984 Poor went on a sabbatical and, with his wife, sailed his yacht to Europe. O'Kelley got tired of flying him back to the U.S. for board meetings and asked Poor to resign, and Poor complied.[172]

"One of the turning points for the company was when Vic Poor went on a cruise," Gistaro said later.[173] "If we had replaced him with someone who was more attuned to what was going on out there technologically I think we would have given Apple a run for their money. After all, we were on the desktop and we had a network. There was a young guy in Canada who is trying to convince us that we should network PCs. He convinced me, but to my fault I did not

---

172 Poor's interview with the Computer History Museum, 2004.
173 Author interview, 2009.

swing enough weight in the company with the technology guys to make them do it. So we missed the PC revolution."

In early 1985, after he joined Image Data Corporation another start-up company, Poor gave an interview with a computer magazine saying he was glad to be out of the computer business and no longer competing with IBM. He also made an off-hand comment that it would be better if the telephone was technologically impossible. If that was the case, humanity would have developed, in place of the telephone, a world-wide network of sophisticated Teletype machines with access to stored records.[174] Ten years later the World Wide Web began to emerge. PCs stood in for the sophisticated Teletype machines, and Web servers offered up stored records.

Sometime after that, apparently in late 1985, Frassanito got a mysterious bill for $600 concerning a Datapoint computer.

"I got on phone and told the fellow that I don't buy computers, I make computers, and any time I need one I go down to the factory and get one," he recalled. "He informed me that the bill was not for the computer but for the monthly service contract from Intelogic Trace. I told him to come by and pick up the machine, and then I bought an IBM PC/AT for a lot less than the service contract would have totaled for a year. It was so incredibly obvious that the Datapoint business model was broken."

By 1987, the year Gistaro left Datapoint because of Edelman's financial manipulations, worldwide microcomputer sales totaled nearly 13 million, and there were nearly 50 million in use. Pundits were talking of a maturing of the market, since yearly sales growth was now "only" in the 30 percent range.[175]

By 1987 the impact that PCs were having on Datapoint's sales could no longer be shrugged off. Revenues were flat in 1986 and 1987, and the firm had lost in those years, respectively, $8.6 million and $52 million. Rather than trying to participate in the runaway PC market, Datapoint put its faith in a series of products usually called the Starship line, as detailed earlier. They were actually symmetric multiprocessors, offering scalable performance across a wide range of products, while requiring the development of only a small number of hardware modules.

Existing customers liked them. They brought in revenue. But they were not PC compatible.

Lock-in Backlash

The problem was not simply that PCs cost less than computers like Datapoint's. After all, the processors of Datapoint and its competitors were premium products backed by premium service, for which you would expect to pay a premium price. The problem is that increasing numbers of computer buyers would not have taken them had they been handed out free. Those buyers were instead acquiring PCs because they needed PCs to run PC software.

After the PC market exploded, millions of people were using some version of MS-DOS (the generic version of the PC operating system, the sales of which made Bill Gates a billionaire.)

---
174   MIS Week, January 16, 1985, interview by Kit Frieden.
175   Sales figures courtesy of Egil Juliussen, Ph.D., president of Computer Industry Almanac Inc.

Thanks to this large market of PC owners, software makers could afford to develop (and massively promote) genuinely attractive packages for the PC, with titles like WordStar, dBase III, WordPerfect, and Lotus 1-2-3. But only in rare cases were versions of such PC software titles offered to run on anything but MS-DOS, since doing so required considerable additional effort by the developers. (Datapoint acquired the rights to Microsoft's Multiplan, a non-graphics spreadsheet, in 1982, and it took two years to port it to RMS.) Consequently, Datapoint users could not run that software—and increasingly resented the fact. But if they stopped using Datapoint computers and switched to PCs—which were cheaper anyway—they could run that software.

As for the computer vendors (like Datapoint) each had its own operating system, and its customers ended up with software that ran only on that operating system. To switch to another vendor meant starting over with new software—a painful experience that most would want to avoid. So when they needed a bigger machine, they often ended up going back to the original vendor—they were locked in.

But PC owners did not feel locked in—they had a large and growing software market they could tap.

"During that period I had every computer company as a client," recalled Amy Wohl, then an office automation consultant. "As for Datapoint, they were a very annoying customer, as they did not want to do anything I told them. I urged them to understand PCs. That was not well received. I urged them to understand the notion that the amount of software that they could get written for their proprietary operating system would be necessarily limited over time. They had to understand that the fact that they had a unique operating system was a limiting factor. Meanwhile, they still had credibility problems, and their products were too expensive for the down-market where credibility was not an issue."

Of course, moving from the Datapoint to the PC environment meant giving up ARC networking. Eventually, networking hardware appeared that offered functionality for the PC that was at least reminiscent of what Datapoint had been offering since 1977. But for new computer buyers who were acquiring a PC to run specific software, and who were unfamiliar with local area networks, history was irrelevant. Many Datapoint users initially remained loyal and resisted the glittering allure of the PC and all those software packages that were being advertised on TV. But the market growth belonged to the PC.

Chapter 18

Death by Irony: 1987-2000

To say it was all downhill after Gistaro's departure (of after Edelman's arrival, or after the debacle, or after any other point) is too simple. Datapoint continued making and marketing new products, but its revenue steadily declined as new generations of increasingly powerful PCs (descendents of the Datapoint 2200) claimed their traditional market.

And management turmoil continued in certain parts of the company, to judge from the experience of Richard Erickson.

In 1986 San Antonio had two competing daily newspapers, the Light and the Express-News. Erickson was a business reporter at the Light, and spent three months researching a history of Datapoint. Even then, he recalled encountering a lot of anger about the fate of the firm, plus nostalgia for the pre-debacle "good old days." As for Datapoint's press relations department, he found that the company was in no hurry to cooperate with him.

But after the series came out in September 1986 he found that Datapoint liked the media attention. And Erickson found himself courted to replace Datapoint's departing press relations manager. He was approached by an executive who reported directly to Edelman rather than to Gistaro, who was still the CEO at that time. After he got to know the executive, Erickson realized that his background was not in media relations, but party and event planning.

"He said they did not like everything that I wrote in the series, but that I had explained it in common language. I refused the offer. They offered to double my salary. I had just gone through a divorce and needed the money, so I swallowed the bait. I regret the decision to this day," he explained later.

"My first day at Datapoint was December 15, 1986. Since I had not been there previously I did not know that they had furloughed the entire company for two weeks without pay, until the end of the year. I was shown the office (by the executive who hired him) and except for him and the security guard I was alone in an eight-story office building. Then my boss said that he was European and that in Europe they don't work during the Christmas vacation, and left for Paris.

"I answered a few phone calls, but I knew nothing about the company or about what was happening there, and there was no one to get any information out of. Basically, I sat there for two weeks. It was the first inkling that I may have made a mistake.

"After that, the first press release I did concerned the fact that Gistaro had left. I was going around visiting various departments to learn what I could about the company. It was obvious that they were looking for positive ways to display the company so Edelman's stock would go up. It became harder after the layoff," Erickson explained.

At the end of February Datapoint laid off 786 employees, of whom 416 were in San Antonio. Erickson remembered that the layoff covered 40 percent of Datapoint's local workforce, and was a major blow to the San Antonio economy. The layoff followed a staggering quarterly loss of $65.86 million.

Gistaro had been replaced by Doris Bencsik, who now had to face the press with the bad news. "Doris was frightened of the media," Erickson recalled. "I wrote her notes for the press conference. I set it up so that she was at a podium at the back of the room and there was a door behind her. I told her that if things got bad she should politely thank them and just slip out the door. I told her to have her facts and figures ready, but the first question would be, 'How do you feel about it?' The electronic media would be after that question. And the press conference opened with a TV reporter asking that very question. So, for several weeks, my stock was golden with Doris. She did not have to slip out the door, either. But the local media really did not know what to ask," he said.

Beyond crisis management, "There was a feeling that we were doomed. Depression was evident throughout the company even before the layoff. There was a feeling that there were no really new products, or innovation, being done. Edelman was not a technical guy and did not care what we produced," he recalled.

As the year dragged on, Erickson's boss suddenly decided that Datapoint's public relations and marketing was "too provincial." Erickson was sent to visit a list of Datapoint offices in London and Northern Europe, while the marketing manager was sent to Paris and Southern Europe. (There was no plan for touring Asia.) At the time 70 percent of Datapoint's revenue was from outside the United States, with the bulk coming from Europe.[176]

"I was not really told why I was being sent. I was given no real message to carry, and had no hard information to give them. Just before I left I was told that there was going to be a change of leadership, but that I could not say anything about it.

"As I was meeting with the head of the London office his secretary came in with the news about Doris being out and Potter being in. (Robert Potter replaced Doris Bencsik as CEO on June 1.) 'You son of a bitch!' exclaimed the manager. I tried to explain that I had been sworn to secrecy and had no hard information anyway.

"Some time after I got back I was told that my department—which was myself and my secretary—would be terminated, and that my boss would take over my duties. That was in October 1987. Whatever was going on that required my departure, it did not seem to be about money. For instance, they had budgeted $50,000 for corporate charity, and I handed it out without needing anyone's approval. I simply gave it away where I thought it would make the company look good, usually with something associated with computers. I gave a branch library $10,000 to buy computers when I noticed that they didn't have any. I just parceled it out to local charities.

"I was packing my personal things in my office on my last day, which was a Friday afternoon, when my boss suddenly asked, 'Who is going to do the annual report?' It suddenly occurred to him that someone needed to go to the printing plant in Dallas and give final approval to the contents of the annual report (then in preparation.) I said that I don't work here anymore. So he agreed to pay me to be a consultant for two days, and I went to the printer with a

---

[176] Datapoint Corporation 1986 Annual Report.

graphics designer.

"Afterwards, I went to dinner with the graphics designer and the head of the printing company, and told them the story. The head of the printing company was astonished. He said that I had had final authority on the contents, and anything I had said to do he would have had to follow through on. Yet I was a potentially disaffected employee. I could have drawn mustaches on the pictures and he would not have stopped me. But I was a good employee and did not do that," Erickson said.

Critical Mass

Michael Fischer, hired in 1985 as Datapoint's principal product architect, remembered the same period as a moderately good time for the engineering department, where there was calm, disciplined, and productive activity. Unlike Erickson, he came to see his decision to join Datapoint as among his best.

Doris Bencsik hired him in late 1984, "To bring Datapoint technology into the 1980s before it becomes the 1990s," he recalled. "She was the person who argued successfully for a $20 million engineering budget in 1985—Edelman wanted it to be $10 million—without which we could not have ever reversed the losses nor developed any new products. Sustaining engineering for what was already in the field cost nearly $10 million per year at that time. She was the person who saw to it that the employees who were still on the payroll were treated well," he recalled.

"Sales of the new products did stop the red ink—Datapoint showed a profit in 1988. What they did not do was permit Datapoint to gain any significant new customers, as opposed to sell more hardware to existing customers. The return to red ink was a direct result of further cuts to the R&D budget that prevented the momentum created by the new products from being maintained, or the holes in the product set to be filled. By 1989 these R&D cuts had reduced the company below critical mass to remain a computer supplier," he said.

As for Edelman, in the three years following his takeover of Datapoint he went after eight more firms: Freuhauf Corp., Ponderosa Inc., Lucky Stores Inc., Burlington Industries Inc., Rexham Corp., Morse Shoe Inc., Foster-Wheeler Corp., and Telex Corp. He had made about $15 million off these projects by the end of 1987. But he was bankrolling these activities using the corporate coffers of Datapoint and Intelogic Trace, making the money unavailable for corporate purposes while it was in play. When he made money with their money, he charged them a commission of 25 percent. Analyst complained that a more normal fee was one or two percent of the amount being managed (regardless of profitability.)[177]

Meanwhile, in a hospital in Bethesda, Maryland, another link to Datapoint's past was broken, when Phil Ray died of cancer on August 13, 1987. He was 52.

Dissolution: Intelogic Trace

When Intelogic Trace was created in July 1985, it had about 2,200 employees and annual

---

[177] Figures and reactions are from press reports in the San Antonio Light on February 2, 1987, and the San Antonio Express News on December 13, 1987.

revenue of about $157 million.

About a year after it was spun off, Intelogic Trace had sold $100 million in junk bonds. Of that, $40 million was earmarked to pay off debts, and $20 million was used to buy various used equipment dealers, since such dealers were thought to represent a good service market. The other $40 million was for investments, presumably to fund Edelman's activities. Payment on those bonds (due twice a year) would create a chronic cash-flow problem that eventually doomed the firm.

Intelogic Trace started out with the domestic[178] Datapoint customer base as a built-in market, and it tried to expand out from there—and encountered enormous competition. It remained in the black at first, but started losing money in 1990. With another bond payment looming, it filed for protection under Chapter 11 of the U.S. Bankruptcy Code in August 1994.

The firm managed to work out a reorganization plan, and emerged from bankruptcy protection on December 8, 1994. The next day, a major customer canceled a service agreement, removing about $500,000 in monthly revenue. Many service agreements had January 1 renewal dates, and as 1995 rolled around a list of other customers canceled. While Intelogic Trace was in bankruptcy, its competitors had been courting them.

It filed for bankruptcy a second time on March 16. There being no hope for a near-term turnaround, it liquidated on April 6, selling its remaining assets for $15.75 million—and then ceased to exist. At the end it had 510 employees, of whom 160 were in San Antonio. In nine years the firm had seen 30 changes in top management, at the level of vice president and above.[179] It lasted ten fiscal years (1985-1994) with cumulative results amounting to a loss of $31 million.

Dissolution: Datapoint

In 1989, Datapoint had its fourth CEO in two and a half years. Near the end of the year it was calculated that in the four fiscal years Edelman had been managing Datapoint its losses had totaled $87.5 million, stockholder equity had shrunk nearly $100 million, and the price of its common stock had fallen from $18 to less than $5 per share.[180]

At that time, Edelman announced that the US has become "reactionary toward a necessary restructuring of American business," and moved to Paris to continue raiding there (without much success, as it developed.) He also stopped talking to reporters.

A year later, Datapoint announced that its headquarters were moving to Paris, since 85 percent of its business was in Europe. Only three executives were involved (other than Edelman.) The CEO at the time resigned rather than make the move, and his replacement was Datapoint's fifth CEO since Edelman took over.

The company had made a profit in 1991 (after two years of heavy losses) but then slid into the red for the next four years. Fiscal 1994 was the worst year, when it lost $95 million on revenue of $173 million, and its auditor expressed doubts about "the company's ability to

---
178 Overseas, service remained under Datapoint's control.
179 San Antonio Express-News, April 16, 1995.
180 Wall Street Journal article, reprinted in the San Antonio Light on December 9, 1989.

continue as a going concern." To make a bond payment, that year it finally sold the tract of land where it had planned to build a massive headquarters complex—a plan that was announced just before the debacle of 1982. It made a profit in fiscal 1996, and a smaller one in fiscal 1997, but then the red ink returned. In fact, it was in the red in seven of its last ten fiscal years.

Dismantling began in 1999, when its final annual report showed revenue of $138,285,000 and a $7,549,000 loss. It still had 639 employees. On May 17, 1999, Datapoint announced the sale of its European operations, which represented 96 percent of its revenues, for $49.5 million. It soon discontinued the MINX, which had produced revenue of only $2.1 million that year. (At this writing another firm was selling the technology, mostly to prisons.) The next year, on May 3, 2000, Datapoint filed for Chapter 11 bankruptcy, and then sold the rest of its European assets.

At this writing, a firm using the name Datapoint remained in business in Europe as a systems integrator for corporate call centers, with headquarters in Brentford, Middlesex, England.

A holding company was left in the US with Datapoint's patent portfolio. The final whimper came in 2005 when it was de-listed by the SEC for not filing regular reports.

An unconnected company called Datapoint USA remained in business in San Antonio, serving the needs of orphaned RMS users.

Final Data

Judged either financially, or by its contributions to technology, there were several Datapoints. In terms of technology, there was the Datapoint of 1969 to 1980, which brought us the desktop computer and the local area network. This was followed by the Datapoint of 1981 to 1985, which had trouble introducing new products and lost its market focus. However, it continued to make money, thanks to the technological lead it inherited from the first Datapoint. Then came the post-Edelman Datapoint, whose technology was gradually strangled by research and development budget cuts as the market shrank.

Financially, there was the Datapoint of 1969 through 1984, which had cumulative revenue of $3.1 billion, and cumulative earnings of $183 million. From 1973 (when it became solidly profitable) through 1984, that Datapoint averaged yearly revenue growth of 38.8 percent. In fact, it averaged 45.4 percent if you stop in 1982, and 50 percent if you stop in 1981.

From 1985 (Edelman's arrival) through 1999 Datapoint's revenues declined by an average of 8.5 percent yearly. Cumulative results for the post-Edelman Datapoint were revenue of $3.76 billion, and a loss of $351.1 million.

Datapoint's lifetime cumulative financial results from 1969 (when it had zero revenue) through its final revenue posting for fiscal 1999 amounted to revenue of $6.85 billion, with a loss of $168.2 million.

From those figures, you'd think Edelman would have gotten out of Datapoint as fast as he could. "I don't know why Edelman didn't sell out," Gistaro said. "I assume a lot of it had to do with ego." The man also proved easily misled about the nature of upcoming products. "He would always over-blow things. He'd say, 'They have this block-buster product they're going to announce.' We'd look at each other and say, 'What the hell is he talking about?' We maybe had an upgrade coming out, but he would say, 'This is it, this is going to revolutionize the entire industry.'" Gistaro recalled.

And, of course, he found he could use Datapoint (and Intelogic Trace) as a piggy bank to back his raiding forays. "He was using Datapoint's capital to do raiding," Gistaro explained. "I think he did all right with it, but he was putting the company's capital at risk in areas that had nothing to do with the company whatsoever."

We'll never know, but perhaps it would have been better if Edelman had announced in that press conference in late March 1985, that he was dismantling Datapoint and selling the pieces. Whatever else would have happened to the segments that survived, they would have been free of Edelman's dilettantism. Without that millstone around their necks, they could have endured or failed on their own merits.

But, frankly, it's highly unlikely that they would have endured. Blaming Edelman is easy, but consider: in the early 1980s, Datapoint's chief competitors in the field of office automation were NBI, CPT, and Lanier. Among computer vendors, its chief rivals were Data General, Wang Labs, Prime, and Digital Equipment Corp. (DEC.)

Their fates:

- NBI—dissolved about 1992. (The author consulted for them just long enough to learn that the name did not stand for Nothing But Initials.)
- CPT—defunct by 1995.
- Lanier—bought by Harris in 1984, later resold to Agfa, and then resold to Ricoh.
- Data General—bought by EMC in 1999.
- Wang Labs—went bankrupt in 1992, later emerged from bankruptcy to be bought by a Dutch firm in 1999.
- Prime—folded in 1992.
- DEC—at one point the world's second largest computer company (after IBM) it tried to enter the PC market four times between 1983 and 1995, each time withdrawing in defeat. It was bought by PC vendor Compaq in 1998, which was itself acquired by Hewlett-Packard in 2002. DEC's founder had opposed advertising, maintaining that good products sold themselves.

The upshot is that the PC tsunami claimed them, too. None survived as enterprises, although fragments of their product lines survived under new ownership. As an independent entity, Datapoint is still in operation and has actually endured longer than any of them.

Even IBM eventually left the PC business, selling that line to a Chinese company in 2005.

Looking Back

"Try not to believe your bullshit so much," said Gistaro, summing up the higher-level lessons offered by Datapoint experience. "Yes, we had this message about integrated technology and multifunctional systems, but the market was telling us, 'I'm not sure that's what I really want. I really need a good word processor, or I really need a good desktop computer that does this or that.' Don't tell them they're full of shit and that you have the answer. Try to listen to the market a little more.

"Also, don't denigrate your competition. Saying you'd be embarrassed to put your name on a Wang or an Apple machine is a little silly when they're both bigger companies than you are.

"This is the hardest thing about the business: understanding that the customers are not innovative," he added. "You can't rely on the customers to tell you what the next great innovation is going to be. But you sure as hell need to rely on the customers to tell you how well received it is. So if you come up with an innovation and you test it in the market and it doesn't play, don't say the market is just stupid. But on the other hand, if you are relying on your customers to tell you what products to build, and how to build them, they won't be innovative.

"I remember a customer in Chicago telling us that we were pretty innovative with this and that. But then he walked into a huge room lined with filing cabinets. There must have been 500 million pieces of paper in there. 'What are you going to do about this?' he asked. There was a need being defined by the marketplace. But how you go about attacking that need, which requires innovation that is over the customer's head? So there has to be marriage between the marketplace and the innovator."

As for Datapoint suffering from having been isolated in San Antonio, Gistaro (who continued living in San Antonio after leaving Datapoint) did not think it was that simple.

"In terms of the ease of managing the company, it was a case of good news and bad news," he recalled. "Family-oriented people would love it because San Antonio is a great family town. Young people, particularly bright young engineers, might rather be in Austin because of its nightlife and intellectual community. In San Antonio, we were the intellectual community, from a technology point of view.

"To some extent we were kind of cut off from the excitement that was happening in California, and that might be the underlying reason that we missed the PC revolution. If you were sitting in a bar every night with guys who are working on Apple computers you might have a different view of the world than if you were in a bar in San Antonio, where the technology gurus were all from Datapoint. There was no one else to sit down at a bar with, unless you wanted to talk about oil drilling. So we had a closed technology community here.[181]

"Before 1982, Datapoint had a great reputation in the industry and was seen as a technology leader. We had the first desktop business computer, the first local area network, and the first video phone. That made it easier to attract good people. And it doesn't snow here," he said.

---

181 Other sources note that San Antonio had a stable technology workforce at a time when Silicon Valley engineers could change jobs by crossing the street.

Of course, it doesn't snow in Silicon Valley, either. Herb Baskin remained there for most of his career, even while developing software for CTC/Datapoint in the 1971-1981 time frame.

"The Datapoint 2200 had many innovative qualities, but you never hear about it since it was not done in Silicon Valley," he said. "If they had been done it in Silicon Valley the people at Datapoint would have been credited with being insightful.

"But they were inadequate at promoting themselves in the technology world. They never established any credibility and standing in the computer science field. Yet, the microprocessor chip and the later ARC system were pioneering things.

"Datapoint played a pivotal role," he concluded.

"Innovator's Dilemma" Postmortem

One of the more highly acclaimed business books to come out since the Datapoint debacle has been, "The Innovator's Dilemma," by Clayton M. Christensen. The book[182] looks at corporate life and death in the face of technological change, and throws a somewhat different light on the fate of Datapoint, and what its management could have done to survive.

Summarizing the book, it states that the common perception of high-tech companies is that they are metaphorically locked in and continual struggle to keep moving uphill in the face of a downhill mudslide. Those who falter and cannot out-climb the mudslide are slowly pulled downhill and out of the market, and often go out of business. Those who get ahead of the mudslide are rewarded with enhanced market share and profits.

But Christensen says that the ability to surmount the mudslide is not what makes the difference between corporate life and death. In fact, fighting the mudslide is a dangerous distraction.

Dealing with the pace of technological progress (the metaphorical mudslide) is what he calls sustaining innovation, and most established companies in a particular market are pretty good at it. Managers at those companies pride themselves on being able to offer better and better products that fulfill the desires of their best and largest customers, especially the customers who buy up-market, high-margin items. That, of course, is where the money is.

But while executing the plans that let a corporation profitably fulfill the desires of its customers, its management may be unknowingly driving that corporation right off a cliff (to switch metaphors.) In fact, the corporations with the most effective and efficient management may hurtle over the edge the fastest.

That is because (while distracted by the metaphorical mudslide) their markets can be suddenly stolen away by new entrants in the field who got there by using that Christensen calls disruptive innovation. Usually the newcomers are marketing a variant of the technology in question, a variant that is initially of no interest to the up-market buyers that the established participants cater to. The newcomers manage to define a viable but small

---

[182] Christensen, Clayton M., "The Innovator's Dilemma: when new technologies cause great firms to fail," Harvard Business School Press, Boston, 1997.

market around a low-cost variation of the technology. Then they move up-market with their disruptive innovation, and eventually offer something desirable to the best customers of the established participants. The customers switch to the disruptive innovation, and the established participants (who have noticed what's going on but have recoiled from moving down-market) see their market evaporate.

Meanwhile, going back to the original mudslide metaphor, Christensen found that getting to the top first was of no particular advantage. Having made a fetish out of it, most established participants in a market are so good at sustaining innovation that they typically offer technology that is more advanced than anything most of the customers are really interested in, or can immediately absorb. Consequently, laggards and me-too companies do fine, as long as they have other ways of keeping the buyers satisfied. Since there are many ways to satisfy a buyer (features, price, convenience, reliability, fashion, software lock-in, etc.) the task is hardly insurmountable.

At least, it's not insurmountable until the new entrants with their disruptive innovation show up and take the market away. At that point the problem does become insurmountable—and irrelevant, because there's no market.

To deal with the arrival of a disruptive technology, Christensen suggests that established participants set up divisions or spin-offs to exploit a new and possibly disruptive innovation, as soon as they identify it. The division should be small, so that the small size of the new market will still seem important to its management. It should have resources separate from the main company, otherwise it probably won't get any resources, since its aims will seem silly if not subversive to the rest of the corporation. The division should plan to identify and develop the market through repeated trials, recognizing that the new market's size is unknown. That will be radically different from that the rest of the corporation will be doing: executing elaborate plans to profitably serve a market of known size.

In Datapoint's case, the sustaining innovation was its progression of processors, with supporting elements such as the ARC local area network. The disruptive innovation was the IBM PC.

By Christensen's standards, Datapoint did the right thing many times. Bringing in a new development group for the 2200 could be seen as the equivalent of setting up a new division to exploit a disruptive technology. The informal group that developed the ARC was another example. Datapoint also set up a division for new telecommunications products, although the commitment of resources was apparently disproportionate to the market. Datapoint could not have known that, but it apparently it did violate Christensen's rule that planning should be limited and tentative when approaching a new market.

But also according to Christensen, Datapoint's processor drought was immaterial, at least initially. Falling behind technologically is rarely a cause of failure, he indicates, as long as desirable customers remain satisfied. Apparently, in Datapoint's case enough of them were, and Datapoint was able to push ahead until its later Starship line of processors came out.

Then, IBM PCs began moving up-market and Datapoint's computer market evaporated.

Here, Christensen's analysis breaks down because the IBM PC was more than a disruptive technology in a specific market, it was a social revolution. It enabled a mass digital information environment, and those who wanted to participate in it needed to get a PC.

So, had Datapoint set up a small PC division in about 1982 (or bought Compaq, which was also located in Texas) that division might have had some success for a while. The more separate it was from Datapoint's main corporate structure (with its up-market attitudes and expectations) the more chances it would have had. But market maturity and consolidation have since produced a situation where any PC vendor that is not in the top five cannot expect significant sales. Scores of vendors were winnowed out in the process, and there is no reason to expect that Datapoint's effort would have fared better than so many others.

Following Christensen's approach, the most promising thing that CTC could have done would have been to keep the intellectual property of the 8008 in 1971 and set up a small division to exploit it. (After all, the chip was the ultimate in disruptive innovations.) Of course, CTC did not have any people with expertise in the microprocessor chip market, and probably did not have the money to pay off Intel and fund a marketing and development effort. But of course, no one else in the world had that expertise either, and if CTC's managers had understood the opportunity they presumably could have raised the necessary money. In fact, it was just the kind of sortie into the unknown that Phil Ray and Gus Roche would have liked.

Finally, when considering Datapoint, Christensen's approach breaks down on another issue: mission. The proceeding analysis is based on the assumption that Datapoint was a technology company, primarily making computers. Certainly, that is what most of the employees thought they were doing. But when seen from Wall Street, Datapoint was a tool for financial exploitation. Thanks to its positive revenue gradients (before the debacle) investors could exploit Datapoint to make money off its rising stock prices. Datapoint's management, meanwhile, would conjure free money from Wall Street by exploiting those investors. After the debacle, Edelman exploited the firm for his own purposes. In that context, it did not matter if Datapoint was making computers or straw brooms. Indeed, at the time, most of its achievements were initially met with incomprehension.

That Datapoint nevertheless accomplished what it did is a tribute to the individuals involved. The direct result is today's digital environment. Datapoint's computer business is gone. The people responsible for its achievements have dispersed to the four winds, or gone to their rewards.

If you seek their monument, look around you.

Chapter 19

Roll Call

The status of some of the entities and individuals behind the Datapoint story during the last half of 2009 (or when last interviewed):

Datapoint lived on in the form of a UK-based systems integrator specializing in corporate call centers.

The unconnected San Antonio-based Datapoint USA continued to support RMS users.

Intelogic Trace was liquidated in 1995.

Intel was a $37.6 billion firm with 84,000 employees, with 80 percent of the microcomputer processor chip market. The last four digits of its main switchboard phone number were 8080.

Microsoft was a $58.4 billion company with 93,000 employees. As with Intel, the last four digits of its main switchboard phone number were 8080.

MINX, the videoconferencing system that emerged from Datapoint, was still marketed by a San Antonio-based company called Vugate, mostly to prisons.

Asher Edelman was running Edelman Arts, an art gallery in New York City.

Gus Roche died as a result of a traffic accident in San Antonio in 1975.

Phil Ray died in 1987.

Intel cofounder Dr. Bob Noyce died in 1990.

Vic Poor was retired in Florida. He had given up flying, but not sailing.

Jack Frassanito was head of JF&A in Houston, Texas, involved in various NASA projects, including the International Space Station.

Harold E. O'Kelley died in 2000.

Doris Bencsik died in 2006.

Dick Norman died in 1992.

Gary Boone, the TI patent-holder, was living in Colorado.

Gerry Cullen worked as in industrial marketing consultant in Austin, Texas.

Ed Gistaro was retired San Antonio, able to live comfortably as a result of businesses he was involved in after leaving Datapoint.

Gerald Mazur was retired in San Antonio.

David Monroe was a principal in the San Antonio-based e-Watch Corporation, selling video surveillance systems.

Herb Baskin left Datapoint in 1981 and later founded a company called General Parametrics, selling equipment that let PCs be used to make presentations. Its market began to suffer after Microsoft released its PowerPoint presentation software, and he sold the firm in 1996. Baskin then became a venture capitalist.

The ARCNET Trade Association had become a Web-only organization and appeared to have only three dues-paying members.

Harry Pyle became a principal software design engineer at Microsoft.

Jonathan Schmidt was executive vice-president of San Antonio-based Perftech (successor of Performance Technology) selling software for broadband Internet providers.

Gordon Peterson was a custom programmer in Dallas.

Hal Feeney was a principal with semiconductor industry consulting firm Pathfinder Research in Cupertino, California.

Ted Hoff worked as a consultant for attorneys involved in patent cases.

Federico Faggin (pronounced fah-Jeen) was retired as the head of Foveon, a maker of image sensors in San Jose, California.

Stan Mazor was retired in Mountain View, California.

Dave Gust was general manager of a San Antonio computer firm.

Michael Fischer worked for a San Antonio high-tech firm. After leaving Datapoint in 1989 he was one of the inventors of Wi-Fi.

Ted Nelson continued to pursue Project Xanadu, and was, among other things, a visiting fellow at Oxford University, and at the University of Southampton.

Appendix A

The Patent Soap Opera

Patent law and intellectual property (IP) rights were almost a non-issue with the original microprocessors. Neither CTC nor Intel initially believed that they were worth patenting, or suitable for patenting. However, two other parties did prosecute patents, and those efforts took on a life of their own.

Meanwhile, Datapoint (with two inventors) later became involved in lucrative patents surrounding videoconferencing inventions. The situation looked promising—but came to grief as a result of a grotesque legal blunder. Finally, a patent suit concerning some of Datapoint's networking inventions similarly came to grief, apparently from its reliance on cut-rate legal representation.

As for the original microprocessor chip, it's clear that, in the early 1970s, computer vendors were just not very excited about patents. Considering that new computer products had a life cycle of about three years before they were obsolete, and it took nearly that long to get a patent, their indifference is hardly mysterious.

"I remember some discussion with Phil and Gus, but they never seemed very interested in patents," recalled Bob McClure, the Dallas-based consultant who originally steered Roche and Ray to their first product. "At the time there were no software patents and relatively few computer hardware patents. There was little industry enthusiasm for patents at that time compared to, say, the pharmaceutical industry, and that is still true to some extent today."

As for the chip that became the 8008, "That spring (of 1970) we had a conference in my office with a patent counsel from Houston," recalled Vic Poor. "We laid out what we were building, looking for patentable things. He didn't see where just putting computer functionality on a single chip was really patentable. All the same functions were from separate parts. Putting them together does not make it patentable. We dropped the issue."

During the design of the 4004, Intel's designers similarly consulted a patent attorney, and were likewise told that the thing was not worth patenting, according to a 1995 interview with Ted Hoff at Stanford University.[183]

Valid or not, such objections only applied to standard "utility" patents, which grant an inventor the exclusive right to exploit an invention for a certain number of years.[184] But there is another class of patent, called design patents, which cover the appearance of a product. They differ from simpler copyright protection in that any copying is an infringement, even if it is not deliberate and the design was arrived at independently. Only non-functional features can be given a design patent—the shape of a gear, for instance, would not be covered, since gears are functional.

On November 27, 1970, CTC applied for a design patent for "the ornamental design for a

---

183 The interview is transcribed at http://www-sul.stanford.edu/depts/hasrg/histsci/silicongenesis/hoff-ntb.html.
184 At the time, a U.S. utility patent had a term of 17 years. It has since been changed to 20 years. Design patents have a term of 14 years.

computer terminal, as shown and described." What was shown and described was a sketch of the enclosure of the Datapoint 2200, from the front, side, top, and oblique, "constructed in accordance with our design." (Only one tape drive was shown, although all production models had two.) It was granted on July 25, 1971 as D224,415, to Ray, Roche, and Frassanito, with CTC as the assignee, with a term of 14 years. (They had previously applied for a design patent for the enclosure of the Datapoint 3300, filed December 17, 1969, and granted as D220,266 on March 23, 1971.)

The Datapoint 2200 design patent is probably that earliest patent that can be directly linked to what is now the PC.

The Boone Patent

As described in Chapter 6, Texas Instruments also set out to make a chip-level version of the Datapoint 2200 processor, at the same time that Intel was trying to do the same thing. TI delivered its chip to CTC about six months before Intel did. CTC could not get it to work, and TI dropped the whole project.

It did not end there, though. TI then went on to patent the chip, filing the application for a "Computing Systems CPU" on August 31, 1971. The patent, number 3,757,306, was granted on September 4, 1973, to TI engineer Gary W. Boone, with TI as the assignee. It is usually referred as the Boone patent. (There were other, similar TI patent applications during that time frame that involved microcontrollers rather than microprocessors, as they included software and memory.)

"I don't know if they (CTC's management) knew anything about the patent—Phil never said anything about it," recalled McClure. "I was only aware of it when TI began to assert it against other vendors in the PC business."

That happened in about 1990. During the pre-trial discovery proceedings a range of people who had been involved in the creation of the 2200 and the 8008 20 years earlier were deposed to give testimony, some multiple times.

Poor recalled being asked why he did not patent the microprocessor back in 1970. The person asking the question was a lawyer from the same firm that, in 1970, had advised him that the chip was not patentable.

"Some of the documentation that TI presented was literally copies of documents that they had received from CTC that they claimed were their own," Poor said. "It was easily traceable. Frassanito wanted everything about the company to be unique, including the type-ball on our Selectric typewriters. So we had a unique typeface. The documents that TI was producing were in our unique font."

It was the content of those documents that galled Intel's Stan Mazor (who was deposed three times), since they showed that the TI chip was built to the initial specification that he wrote for CTC. As proof, he pointed out that TI's specification included an error he made in the initial 8008 specification, which he fixed before the 8008 went to production.

The error, he explained, was that the device was mistakenly designed to jump to an interrupt-handling subroutine when it received an interrupt. (Interrupts are signals from parts of the computer system that are demanding immediate attention from the central processor.) Jumping means that execution is handed off to the code that is contained in the memory address or register that the jump is made to, without any consideration about the future. But a subroutine is supposed to finish its job and then return control to the code that was running before the subroutine was invoked. That means that a return address has to be provided. This is accomplished with an instruction that is referred to as a "call," and it's obviously more complicated than a simple "jump" instruction. The final version of the 8008 correctly "called" the interrupt-handling subroutine, instead of "jumping" to it.

"I believe that the first interrupt would have crashed it," Mazor said. "The TI chip would never have worked even if it had functioned. But that is not an issue when it comes to getting a patent."

Mazor insisted that someone at CTC must have handed his specification over to TI. There would have been nothing to stop them, since no non-disclosure agreement was signed between CTC and Intel.

The TI patent infringement case based on the Boone patent never went to trial. Consequently, what was said in the depositions never became public record. When interviewed for this book, some participants still felt constrained by the witness confidentiality rules. Often, the notes and souvenirs they had provided as trial exhibits still had not been returned, or were assumed to still be covered by non-disclosure restrictions.

Presumably there was some kind of settlement between TI and the PC vendors, but it was never disclosed. However, McClure recalled noticing that, at some point, TI stopped listing the Boone patent in the list of patents that it would assert against infringement.

This may be because TI concluded that the patent would not have survived a trial. Or, as the discovery process dragged on, TI's patent lawyers saw a news item that made them realize that they had a more urgent problem.[185]

Submarine Attack

Basically, the computer industry was stunned when, on July 17, 1990, U.S. patent 4,942,516 was awarded to Gilbert P. Hyatt[186] of California for "single chip integrated circuit computer architecture." In other words, after it had been on the market for almost two decades, a patent on the microprocessor had been awarded to a relatively unknown engineer and consultant. Since the original filing was December 28, 1970, his patent had priority over the Boone patent. It was actually broader than the original Boone patent, since, besides the processor, it claimed the presence of memory and programmed ROM on the chip, amounting to a microcontroller.

The Hyatt patent was what they called a "submarine patent," since it lay undisclosed in the patent office until it suddenly surfaced, with disruptive results. Patent laws (since modified) allowed a patent applicant to keep filing revisions on the original patent. The revisions

---
185 Neither Boone nor the TI patent counsel would agree to be interviewed for this book.
186 Hyatt could not be reached for comment.

would retain the priority date of the original patent, but the filer could add modifications to the patent application to match on-going technological advances. Meanwhile, the existence of the patent remained undisclosed, and vendors that used the technology might suppose that it had passed into the public domain—until the submarine surfaced. Filing patent revisions was an expensive process, but some people made millions using the method.

As for Hyatt, in 1991 Philips NV, the Dutch electronics conglomerate, not only agreed to license several of his patents, including the microcomputer patent, but to act as his agent in getting other firms to pay for licenses.[187] Reportedly, by 1992 Hyatt had already collected $70 million in royalties (although probably not all of it came from the microcomputer parent.)[188]

TI launched a counter-attack and after five years got the Hyatt patent invalidated, in 1996. TI was upheld on appeal in 1998. Basically, the court found that Hyatt's patent application did not include an adequate description of a microprocessor until its 1977 revision, giving priority back to Boone.[189]

At the same time, TI successfully petitioned the Patent Office to turn Boone's microcontroller patent application into a "Statutory Invention Registration" (SIR.) An SIR contains all the information found in a patent (including a full description of the invention) but abandons the right to an actual patent (and thus the right to claim royalties) on the invention. Basically, it establishes the applicant (who has to pay all the usual patent filing and processing fees) as the inventor, while putting the invention in the public domain.

Barring the successful appeal by Hyatt (and at this writing none had materialized) that is where the issue stands.

Hyatt last appeared in the media in 2002, when he was the subject of a lawsuit between California and Nevada. California felt that some of Hyatt's millions were subject to California taxation. Nevada (where Hyatt had become a resident of Las Vegas) agreed that Hyatt should be allowed to sue California for invading his privacy during the process of investigating his residency status.[190]

The Videoconferencing Patent Debacle

As explained in Chapter 15, David Monroe decided to leave Datapoint after running afoul of Harold E. O'Kelley just before the start of its financial debacle in 1982. His intention was to form a company to exploit the videoconferencing technology that he and Frassanito had been developing since 1974.

They had had a working system going by 1979. The TV signal remained analog (unlike much later PC-based systems) but the system used a "frequency agile" monitor that could switch between data, videoconferencing, and broadcast TV. Each unit also contained a camera and

---

187 See http://query.nytimes.com/gst/fullpage.html?sec=technology&res=9D0CE7DF1531F934A35752C1A96795 8260&scp=1&sq=Gilbert%20P.%20Hyatt&st=cse, accessed September 14, 2009.
188 See http://www.ipo.org/AM/Template.cfm?Section=Search&template=/CM/HTMLDisplay. cfm&ContentID=10692, accessed September 14, 2009.
189 See http://www.ipo.org/AM/Template.cfm?Section=19981&Template=/CM/ContentDisplay. cfm&ContentID=3674, accessed September 10, 2009.
190 See http://query.nytimes.com/gst/fullpage.html?res=9B07E0D7163DF935A25753C1A9649C8B63&scp=5&sq= Gilbert%20P.%20Hyatt&st=cse, accessed September 14, 2009.

microphone. By the early 1980s they had advanced to the point where the system could automatically show the face of whoever was speaking in a multi-person conference.

As he began setting up the new venture, he was unexpectedly approached by Vic Poor, who told him that Datapoint was working to exploit the videoconferencing technology, and would fight Monroe in court if he sought to develop and market the same technology.

By right of the terms of the employment contract he had signed after returning to Datapoint after Mnemonics failed, which named him as the owner of the videoconferencing technology with permission to work on it, Monroe assumed that the IP situation was clear. But the attorneys he consulted said he was not in a position to sue Datapoint, since Datapoint had not yet made or sold any videoconferencing systems. Therefore, there were no damages to claim and (most especially) no contingency fees for the lawyers. Hiring lawyers on a retainer basis to stop Datapoint would have cost at least a million dollars, Monroe was told.

So he took the advice of the lawyers and pursued another course while waiting for Datapoint to do something that could be the basis of a lawsuit. He, Frassanito, and Gerald Cullen partnered to form Image Data Corp., which made equipment that transmitted pictures over ordinary phone lines using a combination of TV technology, fax technology, and digital compression.

As described in Chapter 16, after Asher Edelman took control of Datapoint in 1985 he visited the company and abruptly decided not to liquidate it because he was impressed by what he saw. This included a videoconferencing system that was under development, based on Monroe's work. Datapoint soon brought it to market under the name MINX.

Monroe thereupon did sue Datapoint for damages. His old employment contract prevailed, and Datapoint settled by splitting the ownership of the patents with Monroe and Frassanito. They subsequently joined a 1993 patent infringement lawsuit that Datapoint brought against PictureTel Corp.[191] of Massachusetts, which had become the leading vendor of long-distance videoconferencing systems.

Datapoint's lawyers claimed damages of at least $200 million against PictureTel. Half of that (minus the lawyers' share) would have been gone to Monroe and Frassanito.

But the month-long jury trial in Dallas in March and April of 1998 was a disaster for them. As Monroe explained, when Datapoint brought its videoconferencing system to market in 1985, it discovered that someone else was already using the name MINX. At that point Datapoint had had the choice of renaming the product and reprinting all its product literature, or suing the other party. They choose the second route and pulled a date out of the air for Datapoint's earliest use of the word MINX. Conveniently, the date pre-dated the other party's use of the word, so the other party backed down and Datapoint could go ahead with the product launch.

However, under patent law, a patent applicant has a year after the initial disclosure of a technology to file a patent for that technology. In the 1998 trial, the court chose to use Datapoint's claimed first use of the word MINX in the 1985 case as the date of the initial

---

191 PictureTel was acquired by Polycom Inc. of California in 2001.

disclosure. However, that date came more than a year before the filing of the patents in question, and therefore invalidated them, Monroe explained.

Datapoint's attempt at an appeal was defeated a year later.

The ARCNETplus Patent "Embarrassment"

As mentioned in Chapter 13, in 1989 Datapoint came out with an enhanced version of the ARCNET local area networking protocol called ARCNETplus. A given port on an ARCNETplus hub could run at the original ARCNET speed of 2.5 megabits, or at an enhanced speed of 20 megabits. Several patents were filed concerning the workings of ARCNETplus, including #5,008,879, "LAN with inter-operative multiple operational capabilities," otherwise known as the 879 patent. Any LAN using multiple transmission rates would use the concepts of the 879 patent, recalled Michael Fischer, one of its co-inventors.

In 1995 the so-called Fast Ethernet protocol was standardized, defining a version of Ethernet that ran at 100 megabits but was backward compatible with existing 10-megabit Ethernet ports. A given network could have a mix of older 10-megabit nodes and newer dual-speed 10 or 100-megabit nodes. The market switched to Fast Ethernet fairly rapidly, and by 1998 annual hardware sales already amounted to $7.6 billion. Patent royalties (usually in the 3 to 8 percent range) would there have amounted to a significant sum, Fischer recalled.

Datapoint's position was that Fast Ethernet clearly violated at least the 879 patent, and filed suits against multiple Ethernet hardware manufacturers in 1996. These were consolidated and resulted in a court trial in January 1998 in Manhattan.

The fact that the trial was held in Manhattan was Fischer's first clue that things were not right. The Federal Eastern District of Texas was famous for being favorable to inventors in patent cases due to its strict compliance with certain procedural rules that had the effect of keeping well-heeled litigants from manipulating the process. Datapoint was eligible to file in the Eastern District of Texas, but its New York lawyers filed in New York for their own convenience.

Then the hearing started and Fischer found that Datapoint's lawyers were obviously unprepared, both in their arguments and in their cross-examination of witnesses. He felt embarrassed to be sitting at the same table with them, he later recalled.

The judge (actually a "special master" with patent experience) later ruled that Fast Ethernet did not violate the Datapoint patents because, with Fast Ethernet, each separate connection between a port on a machine and a port on a hub is a separate network running at one speed. Logically, of course, the network extends through the port on the hub and then out its other ports, which may be running a different speeds, but the judge did not see it that way, or simply did not understand the arguments that Datapoint's lawyers tried to present.

Datapoint appealed, meanwhile filing more suits against makers of Apple's FireWire network and its variants. However, the original decision in favor of Fast Ethernet was upheld in 2002. (By then Datapoint was bankrupt and the matter was being pursued by Edelman's successor company, Dynacore.) In 2003 the court decided that the Fast Ethernet decision applied in the FireWire cases, and that suit was dismissed.

And there it ended.

Appendix B

Timeline of Events

1929
November 11: Austin Oliver "Gus" Roche is born in Brooklyn, New York.

1935
Unknown: Jon Philip Ray is born, possibly in Austin, Texas (according to his obituary) or in Shreveport, Louisiana (according to a 1970 Dun & Bradstreet report concerning CTC.)

1967
Both Roche and Ray live in Florida. They are both NASA engineers.

Unknown: Vic Poor arranges a meeting with Plantronics executives for Roche and Ray.

Summer: Phil Ray and Charles Skelton visit Bob McClure in Dallas. He advises them to make a "glass Teletype."

1968
Spring: Gerald Mazur raises money to start CTC in San Antonio.

March: Jack Frassanito joins Raymond Loewy and works on the Skylab project.

July 6: CTC incorporates under Texas law.

July 18: Intel incorporates in California.

July 31: Fiscal year begins for Computer Terminal Corporation. They operate out of 142 West Rhapsody.

August 23: A CTC letter signed by Ray is sent to Raymond Loewy inquiring about industrial design services.

1969
January 2: Breadboard of the Datapoint 3300 works for the first time.

May 14: The Datapoint 3300 is introduced at the Joint Spring Computer Conference in Boston.

June 20: Busicom executives including Masatoshi Shima arrive at Intel to discuss a calculator.

July 27: CTC has orders for 876 terminals over the next 18 months worth $2.9 million.

August: CTC goes public.

September 21: CTC ships first products for revenue.

September 25: CTC has orders for 1,405 terminals and 141 magnetic tape units, worth more than $5 million.

September: Shima visits Intel about the 4004 project, and assumed he would check the design on his return in April. Mazor leaves Fairchild for Intel and is assigned to the 4004 project.

Thanksgiving: Poor and Pyle design the Datapoint 2200 instruction set.

November: CTC's new 21,000-square-foot facilities are finished at Wurzbach and I-10, then a rural area.

December: Masatoshi Shima submits final functional specifications for the 4004 to Intel.

Christmas Season: Vic Poor meets with Intel executives and agrees to a single-chip processor.

1970

February: Intel signs 4004 development contact with Busicom. Poor sends Mazor the Datapoint 2200 instruction set.

Unknown: Ray convinces Intel to take on the CPU chip project. Texas Instruments also agrees to make the chip. Poor finishes his design without waiting for either chip.

March 9: Hal Feeney begins work at Intel, and his first assignment is the 1201 chip for CTC.

April 3: Federico Faggin is hired at Intel for the 4004 project, and finds that no progress had been made.

April 4: Shima returns to Intel and discovers the lack of progress on the 4004. He stays and helps Faggin with the design.

April: The first prototypes of the Datapoint 2200 are sent to the American Bankers Association convention in San Francisco.

June 21: Penn Central Railroad declares bankruptcy. That day Ray, Roche, and Mazur were in New York fruitlessly trying to raise money.

July: Intel stops work on the 1201 chip.

November: The Datapoint 2200 is announced.

November 27: A design patent, good for 14 years, is filed for the appearance of the Datapoint 2200.

December 28: Gilbert Hyatt files for a patent for a computer on a chip. The patent is not granted until 1990, and is later invalidated.

Christmas Season: The first 4004 fails—one of the masking layers had been omitted.

1971

January 4: A Japanese customer expresses interest in the 1201 chip, and development resumes.

Mid-January: Second attempt at making the 4004 succeeds.

February 24: Texas Instruments memo describing their Datapoint 2200 processor chip is written, longhand.

April 7: First Datapoint 2200 installation takes place at the Pillsbury headquarters in Minneapolis.

June 7: TI announces their version of the 2200 chip in Electronics, implying it will be used in the Datapoint 2200 Version II.

Summer: TI delivers its version of the Datapoint 2200 CPU chip. It is unreliable and CTC rejects it.

Early July: TRW agrees to invest in CTC in return for handling CTC's overseas manufacturing and sales.

July: Mike Faherty arrives as controller.

August 31: Intel engineer Gary Boone files a patent for the TI CPU chip that was derived from the Datapoint 2200 project, resulting in a series of patents issued in 1978.

September 1: Mazur resigns as CTC's chairman of the board.

Unknown: Intel delivers the 8008 chip to CTC. It is inferior to the second generation Datapoint 2200, and CTC rejects it.

November 15: Intel advertises the 4004.

December: Faherty issues his report to the investors: let CTC go under (and face lawsuits) or invest enough to launch the Datapoint 2200 Version II.

1972

January 12: CTC avoids bankruptcy minutes before the annual stockholders meeting.

April 15: Intel announces the 1201 chip as the 8008.

April 30: CTC nine-month financial report mentions the Datapoint 2200 Version II.

May: Intel commits to developing the 8080, about six months after Faggin suggested it.

July 25: Design patent 224,415 is issued to Ray, Roche, and Frassanito for the appearance of the Datapoint 2200.

December 7: CTC is renamed Datapoint Corporation.

1973

March 11: Former engineer and Harris Corp. executive Harold E. O'Kelley becomes president and CEO of Datapoint.

July: TRW agreement is renegotiated so that TRW retains international sales but not manufacturing. Ed Gistaro becomes head of marketing.

October 31: So far Datapoint has sold 3,071 Datapoint 2200s.

1974

April: Intel markets the 8080 for $360.

July 17: Ray and Roche leave Datapoint to start Mnemonics.

October 31: Faggin leaves Intel to form Zilog.

Unknown: Frassanito leaves Datapoint to found Mnemonics and later John Frassanito & Associates (JF&A).

1975

January: First sales of the Datapoint 5500 "mini-mainframe."

February 12: Gus Roche is injured in a one-car accident at about 1:15 am.

February 15: Gus Roche dies of his injuries.

Thereafter: Mnemonics is dismantled.

1976

February: Faggin's new venture, Zilog, has 11 employees.

Summer: Development begins on the Datapoint ARC local area network.

1977

January 17: Datapoint buys Amcomp, a maker of tapes and drives, for $2 million.

February: First sale of an Infoswitch, Datapoint's long distance phone management system.

March 13: Michael Faherty resigns.

April 4: Datapoint starts trading on the NYSE with the DPT symbol.

June 13: Datapoint 6600 released at the National Computer Conference, sporting the new 16k RAM chips and supporting 24 terminals.

September: The first ARC network is installed at Chase Manhattan Bank in New York City.

September 20: Datapoint releases an ACD (automatic call distributor) and other hardware for corporate call centers.

October: Datapoint unveils the 1500, based on the Z80.

December 1: ARC is unveiled at a press conference in New York City.

1980

(Throughout): Intel launches Operation Crush to defeat the threat of the Motorola 68000 to the Intel 8086.

September 23: Datapoint buys Inforex.

November 14: Datapoint introduces the 8800 and RMS. The 8800 supports an amazing one megabyte of RAM and a gigabyte of online storage.

End of Year: Faggin cashes out of Zilog and puts his money in the bank, making interest of $2,000 per day.

1981

February 13: Datapoint buy's TRW's international sales and distribution network of Datapoint products.

April 2: ISX announced at a New York press conference.

August 12: IBM unveils the PC.

September 9: Datapoint introduced the 8600 desktop computer with a detachable keyboard, amber screen, 256K, and an integrated ARC interface that eliminates the need for a RIM box.

October 12: Ground is broken for a new Datapoint corporate headquarters on a 148-acre site on I-10 between DeZavala and Hausman roads. The firm's 3,400 San Antonio employees at that time worked out of 36 different buildings.

November 16: Datapoint unveils a color graphics system, a laser printer, and a fax interface.

November 24: Datapoint lays off 150 people.

1982

January: Datapoint executives forecast income of 66 cents per share for the fiscal 2Q about to end January 31, representing growth of 10 percent rather than the 35 percent income growth rate Datapoint investors were accustomed to. The stock price drifts down to about $50, leaving a still unusually high price-earnings ratio of about 25, implying that investors continue believing in Datapoint's future growth.

Feb. 1: Datapoint's stock price is $49.38, and 97,000 shares were traded.

Feb. 2, Tuesday: Datapoint announces that earnings for the second fiscal quarter ending Jan. 31 would be about 56 instead of the previously predicted 66 cents per share. The stock falls to $41.00.

Feb. 3: $36.25

Feb. 4: $37.25

Feb. 5: $34.88

Feb. 8: $33.25

Feb. 9: $33.75

Feb. 10: $31.50

Feb. 18: Datapoint's management calls a meeting with analysts to reassure them. It doesn't work.

February 19: $25.63

March 24: Gistaro becomes new COO.

March 26: Datapoint freezes wages and begins layoffs (230 in San Antonio, 350 in Waco) and delays construction on its new headquarters.

April 8: $22.12, the price falling $3.25 following an announcement that the firm was reducing its field sales force

April 30 (Black Friday): $16.75, down $4.87, on rumors that Datapoint would announce a loss. Trading of Datapoint stock had to be suspended for 3.25 hours.

May 3 (Black Monday): $13.13, after trading again had to be suspended temporarily. The Wall Street Journal runs a short article reporting that Datapoint VARs had been pressured to write shaky orders to keep the sales numbers up.

May 5: Datapoint admits loss of $22.9 million, and was wiping away $105.9 million in suspicious orders. But new orders worth $117 million remain on the books.

May 13 (Black Thursday): Four executives are fired and one is demoted.

May 27: The Wall Street Journal runs a lengthy front-page story on the misdeeds of the Datapoint sales force.

October 29: The Wall Street Journal reports that Datapoint stockholders were receiving proxy material reporting an SEC probe.

1983

Datapoint makes #447 on the Fortune 500. It's the only San Antonio company on the list.

May 9: The 5,000th ARC is sold (4,000 domestically, 1,000 overseas.)

May 16: Datapoint sells its Communications Management Products division.

1984

June 5: Datapoint goes public with the ARCNET protocol at a NYC press conference.

June 18: Without admitting wrongdoing, Datapoint and a former corporate vice president consent to an order by a federal judge barring them from future violations. The SEC said that Datapoint had overstated revenue in fiscal 1981 by $22.1 million.

September: Vic Poor resigns from Datapoint.

November: Datapoint stock rises to $14, indicating that someone is attempting to buy large blocks of stock.

December 9: SEC Form 13-D is filed by a group led by Asher Edelman, announcing they own more than 5 percent of Datapoint stock.

December 19: 240 employees are laid off in Fort Worth.

Throughout: Frassanito breaks ties with Datapoint and returns to Houston to work for the space program.

1985

January 11: Edelman offers $23 per share for the 18.1 million Datapoint shares he doesn't already own.

January 14: Datapoint rejects Edelman's takeover bid.

January 25: Datapoint complains that the takeover bid has exacerbated its financial woes and lays off 659 people.

February 4: Edelman sues Datapoint to overturn its suddenly changed corporate bylaws which would require that solicitation offers take 90 days to complete.

March 2: Datapoint sells Inforex for $12 million plus the assumption of $21.5 million in Inforex debentures.

March 6: Edelman wins a lawsuit permitting him to mail proxies to Datapoint shareholders.

March 15: Edelman effectively takes over Datapoint at a contract signing in New York City. O'Kelley quits, and retires.

March 29: Edelman announces that he will not liquidate Datapoint. The stock falls $3.

May: Datapoint announces that its service division will be spun off into a separate company. Its stock rises to $15.

May 2: Datapoint lays off 310 people.

May 13: John C. Butler is hired as Datapoint's new COO.

June 10: Butler is fired.

July 1: Datapoint announces the name "Intelogic Trace" for the service spin-off.

July 30: Datapoint announces MINX.

September 3: S&P puts Datapoint on its credit watch list.

1986

Unknown: Edelman goes on to buy 17 percent of Datapoint's stock.

July: Intelogic Trace sells junk bonds worth $100 million.

Autumn: Edelman sets out to convert 8 million shares of Datapoint common stock into 2 million shares of preferred stock paying a 13 percent dividend. The SEC filing says that company directors and executive officers would convert only an insignificant amount of their holdings, but Edelman and his backers convert all their shares

1987

January 15: Gistaro resigns.

February 28: Datapoint lays off 786 people.

August: ARCNET Trade Association founded.

August 13: Jon Philip "Phil" Ray dies.

September 3: Doris Bencsik resigns as CEO, and as a board member, severing all ties with Datapoint. Robert Potter replaces her.

1989

Unspecified: Edelman moves to Paris, stops giving interviews.

September: Edelman uses working capital from other firms under his control to boost his control of Datapoint from 10 percent to 40 percent in order to avert a hostile takeover by Martin Ackerman. Intelogic Trace contributed $15 million.

September: ARCNETplus announced, pushing ARCNET speeds to 20 megabits. Delays in getting it to market prevent it from having any impact.

October: Datapoint launches an unsuccessful ad campaign intended to make it a household word, using the slogan "Datapoint, your link to computing power."

1990

March 11: Datapoint has lost $52 million, laid off 588 people, and changed top management three times in 18 months.

December 19: Datapoint moves its headquarters to Paris, now that 80 percent of its revenues come from Europe. The move only involves three executives other than Edelman.

1994

August 1: Datapoint imposes a four-day work week as a cost-cutting measure.

August 5: Intelogic Trace files for Chapter 11 bankruptcy. It had 720 employees, 220 of them in San Antonio.

November 22: Datapoint sells the vacant 148-tract that had been intended for its headquarters.

December 8: Intelogic Trace emerges from bankruptcy.

December 9: MCI cancels a huge maintenance contract with Intelogic Trace, removing revenue of $500,000 per month.

1995

March 16: Other clients having failed to renew their service contracts, Intelogic Trace re-enters bankruptcy.

April 6: Intelogic Trace is liquidated and its remaining assets are sold for $15.75 million. In the end it has 510 employees, 160 in San Antonio.

1996

Datapoint sells its European based Automotive Dealer Management Systems business.

1998

January: Datapoint loses patent infringement case against Fast Ethernet.

April: Court finds for PictureTel against Datapoint et al.

June: Patent court finds against Hyatt and for Boone.

1999

May 17: Datapoint sells its European operations, which represent 96 percent of its revenues, for $49.5 million

2000

May 3: Datapoint files for Chapter 11 bankruptcy.

June 19: Datapoint sells its name and various operations to its European subsidiary for $49.3 million. The U.S. remnant changed its name to Dynacore Holdings Corp. and pursues patent infringement suits based on the Datapoint networking patents.

December 8: Harold E. O'Kelley dies in Indialantic, Florida.

2003

February: Dynacore takes over the CattleSale Company, which had been Edelman's original flagship operation.

February: Dynacore loses patent infringement case (inherited from Datapoint) against FireWire.

2005

August: CattleSale was de-listed by the SEC for not filing regularly.

2008

June 23: Technology industry analyst firm Gartner Inc. said that there were one billion PCs in use worldwide. The number was growing 12 percent yearly and should surpass 2 billion in early 2014.

Datapoint Time Line of Invention

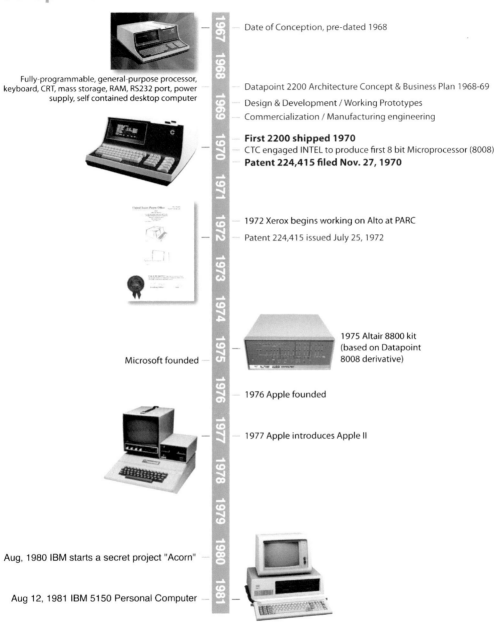

- 1967 — Date of Conception, pre-dated 1968
- Fully-programmable, general-purpose processor, keyboard, CRT, mass storage, RAM, RS232 port, power supply, self contained desktop computer
- 1968 — Datapoint 2200 Architecture Concept & Business Plan 1968-69
- 1969 — Design & Development / Working Prototypes
- Commercialization / Manufacturing engineering
- 1970 — **First 2200 shipped 1970**
- CTC engaged INTEL to produce first 8 bit Microprocessor (8008)
- **Patent 224,415 filed Nov. 27, 1970**
- 1972 Xerox begins working on Alto at PARC
- Patent 224,415 issued July 25, 1972
- 1975 Altair 8800 kit (based on Datapoint 8008 derivative)
- Microsoft founded
- 1976 Apple founded
- 1977 Apple introduces Apple II
- Aug, 1980 IBM starts a secret project "Acorn"
- Aug 12, 1981 IBM 5150 Personal Computer

Appendix C

Datapoint Financial Results

Year	Revenue	Profit (Loss) Before Taxes	Number of Employees	Fortune 500 Position
1969	0	($688,000)	70	
1970	$3,847,000	($1,216,557)	200	
1971	$3,098,000	($3,749,969)	274	
1972	$5,410,000	($2,220,000)	305	
1973	$18,645,000	$1,957,000	608	
1974	$34,063,000	$3,423,000	1,113	
1975	$46,890,000	$4,617,000	1,286	
1976	$72,050,000	$7,834,000	1,921	
1977	$103,023,000	$11,502,000	2,732	
1978	$162,261,000	$15,278,000	3,889	
1979	$232,101,000	$25,246,000	5,066	
1980	$318,826,000	$33,478,000	5,939	
1981	$449,490,000	$48,761,000	7,915	
1982	$508,486,000	$2,405,000	8,822	
1983	$540,192,000	$8,077,000	8,914	447
1984	$600,154,000	$28,182,000	8,413	440
1985	$520,168,000	($48,000,000)	5,993	416
1986	$325,227,000	($8,558,000)	3,612	458
1987	$312,000,000	($58,000,000)	2,749	
1988	$331,000,000	$8,000,000	2,693	
1989	$313,000,000	($29,200,000)	2,451	
1990	$267,311,000	($88,812,000)	1,810	
1991	$265,479,000	$5,335,000	1,741	
1992	$255,243,000	($10,409,000)	1,777	
1993	$208,344,000	($11,859,000)	1,528	
1994	$172,936,000	($94,765,000)	1,444	
1995	$174,901,000	($28,343,000)	991	
1996	$179,541,000	$19,342,000	705	
1997	$142,121,000	$2,383,000	641	
1998	$151,445,000	($669,000)	652	
1999	$138,285,000	($7,549,000)	639	

Appendix D

Processor Ratings

Michael Fischer, who was Datapoint's principal product architect from 1985 to 1989, researched the following chart and graph, comparing each of the principal Datapoint processors until 1990. The performance of each (calculated from numerous factors) is rated in terms of the Datapoint 2200 Version II being equal to one. Also calculated are the relative performance of the first Intel (and Zilog) processor chips, and the first several generations of PC systems.

The Datapoint processors and the PCs can be considered a direct comparison, as they both represent full systems. Comparing the processor chips of these systems is less direct, but the chips are included for historical purposes.

Please note that the vertical scale in the chart is not linear, with each grade being 10 times greater than the one below it. (A linier version insert of 1970 to 1990 shows the 3X performance advantage to scale)

The results clearly show that Datapoint's processors hit a plateau in 1975, with only incremental improvement until 1985. (This is the "processor drought" mentioned in Chapter 17.) But the processor chips, and later the PC systems, quickly began conforming to Moore's Law, and thereupon showed continuous, dramatic improvements.

It's also clear that Datapoint's processors broke away from the plateau after 1985. Using multiple processors, they were able to maintain growth gradients greatly exceeding those of the PCs and chips. For instance, the 7800 used two 80386 chips, the 7950 used four 80386 chips, the 7850 used two 80486 chips, and the 7960 used four 80486 chips. The latter had almost three times the relative performance of a 486 PC. But the Datapoint processors were not PCs, and did not run PC software (a management choice not a technical limitation). And even with this far superior processor performance the management wasn't able to reinvent the business model and ultimately failed even with these potentially disruptive "game changing" breakthroughs.

Model	Year	Relative Performance	CPU Width (bits)	Memory Width (bits)	Max Memory (bytes)	Clock Rate (MHz)
Datapoint Systems						
2200 V. I	1970	0.02	8	8	8K	0.125
2200 V. II	1972	1.0	8	8	16K	2
5500	1975	3.1	8/16	8	64K	5
6600	1977	4.2	16	8	256K	6.67
1500	1977	3.5	8	8	64K	4
1800	1978	1.2	8	8	64K	2.5
3800	1979	1.3	8	8	128K	2.5
8800	1980	5.3	16	16	1M	5
8600	1981	4.9	16	16	512K	4
8400	1984	5.1	16	16	1M	6
8850	1985	5.3	16	16	1M	5
7600	1986	5.1	16	16	2M	6
7900	1986	35.3	16	32	8M	10
8865	1987	9.8	16	16	4M	8
7800	1987	78.5	32	32	8M	16
7700	1988	43.6	32	32	8M	16
7950	1988	161	32	32	32M	20
7850	1990	367	32	32	16M	25
7960	1990	713	32	32	64M	30
Intel and Zilog Processor Chips						
8008	1972	0.26	8	8	16K	0.8
8080	1974	1.5	8	8	64K	2
Z80	1976	3.5	8	8	128K	4
8086	1978	4.8	16	16	1M	5
80286	1982	4.8	16	16	16M	8
80386 DX	1986	48.3	32	32	4G	16
80486 DX	1990	231	32	32	4G	33
PC Systems						
PC	1981	3.7	16	8	640K	4.77
PC/XT	1983	3.7	16	8	640K	4.77
PC/AT (6MHz)	1984	11.3	16	16	4M	6
PC/AT (8MHz)	1985	14.8	16	16	6M	8
386 PC	1987	43.6	32	32	8M	16
486 PC	1990	209	32	32	16M	33

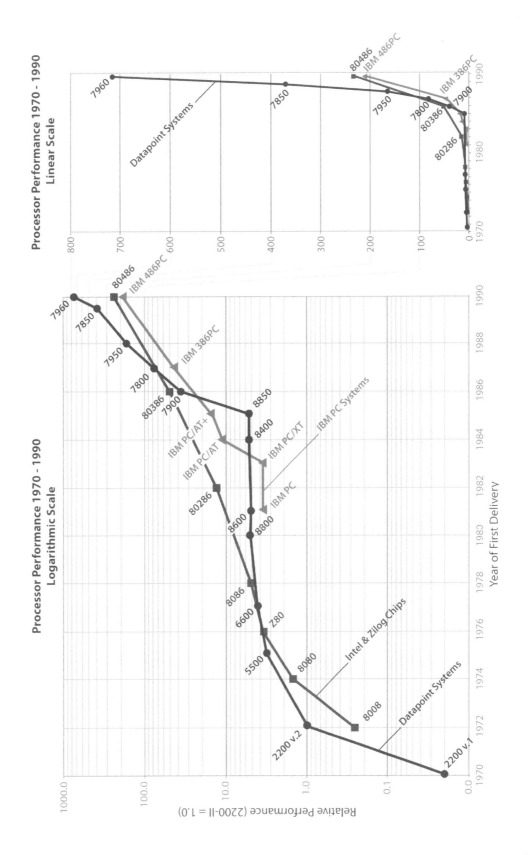

Appendix E

Computer Terminal Corporation Business Plan

PROPOSAL TO ORGANIZE

COMPUTER TERMINAL CORPORATION

TABLE OF CONTENTS

    I. Introduction
   II. Markets and Products
  III. Personnel
   IV. Financial Projections
    V. Conclusion

       Appendix A   IDS Annual Report for 1963

I. Introduction

It has been predicted by Bell Labs that by 1975 there will be more computers communicating with other computers than there are currently people communicating with other people.

Regardless of the accuracy of this prediction, the coming "Information Explosion" is self evident if you consider the recent short trading days by both the New York Stock Exchange and the American Exchange. Man will adapt to the "Information Explosion" by applying the technologies developed in the fifties and early sixties which were applied to the missile and space vehicles of that period. Specifically these are the application of basic information theory developed by Shannon, Phister and other early pioneers by using solid state integrated circuits and memory techniques. Thus to profitably enter the market in the new era, it is necessary that a company have personnel skilled in all disciplines utilized in this technology.

The "Information Explosion" will require sophisticated hardware to meet the increasing volume of information. This hardware is listed below:

1. Digital Computers.
2. Input-output equipment.
3. Terminal equipment.

The digital computer industry can be broken into two categories. These are large general purpose computer and the small general purpose computer. The large computer industry is dominated by such firms as IBM, Control Data, RCA, and Honeywell, while the small computer industry is represented by such firms as Scientific Data Systems, and Digital Equipment Company. Both of these fields are well defined markets in which the need is being met by well established firms. Any effort to penetrate this market would be both costly and difficult thus this proposal is not intended to imply any effort to enter these markets.

Input-output equipments are comprised of such items as card sorters and readers, tape readers, tape and card punches, line printers, digital plotters and other similar devices. A few companies prominent in these areas are Calcomp, Ampex, Burroughs, Frieden and others. While input-output equipments are used in terminal systems, there is however, a whole field of terminal equipment which is necessary for an information processing system to function for both "Real Time" and "Time Shared" systems.

Terminal equipment is comprised of those type devices and systems which enable computers to communicate with one another, and in addition, to permit a remote data transmission source to operate a display system, actuate a control system or simply enter data into the memory of another computer for further processing. The initial equipment to be manufactured by Computer Terminal Corporation will be used to provide translation of standard teletype data in order to generate a video signal for display on a cathode ray tube.

Thus this proposal is directed toward the organization of a company devoted to the development and marketing of terminal data handling systems in order to fulfill a current specific need as well as those markets which will develop in the coming years.

Computer Terminal Corporation can meet this need and prove to be an extremely profitable investment for those with sufficient vision and courage to make an initial purchase of common stock or debentures.

II. Markets and Products

Ideally the product selected for initial development would satisfy the following conditions:

1. <u>Minimum Market Risk</u> - The product should satisfy a specific and immediate requirement plus being technically and cost competitive.

2. <u>Minimum Development Risk</u> - The product should be simple enough to allow development by a small number of qualified personnel in a short time period.

3. <u>Self-sufficent</u> - The product should satisfy a particular requirement without the need of additional support equipment. That is, it should be a "stand-alone" item.

4. <u>Related to Future Goals</u> - The product should be a stepping-stone to the achievement of the long term corporate goal. It should fit in as a portion of the computer terminal system to eventually be developed by the company.

The product chosen to satisfy each of these conditions is the TV compatable display unit. This device accepts alpha-numeric (letters and figures) data in teletype or computer code and converts the data to visual display form. The display is presented on a cathode ray tube (TV type).

There is a tremendous general need for an inexpensive device of this type, but the immediate requirement is for the presentation of press-wire news data over CATV networks. Presently this requirement is being temporarily satisfied by the use of a hard-copy teletype printer and a TV camera. However, the news service companies (AP, UPI, etc.) are very concerned with the maintenance and down-time cost of such a system. The proposed product involves no moving parts and offers years of trouble free service.

Once developed, this product fits perfectly into many other applications. For example:

1. Computer terminal soft copy displays.
2. Airline reservation and schedule type displays.
3. Stock market and commodities market quotations.
4. Inventory and stock data in both manufacturing and warehousing.

The market for an inexpensive computer terminal display alone is practically unlimited.

Subsequent products to be developed would be directed toward the completion of the computer terminal system. The next product following the Display System would be a low speed, low cost "Modem" and an interface device to function with the Bell System Data Phone.

The annual market for computer terminal equipment is shown in Figure 1.* Figure 1 indicates that we are just entering the period of maximum market growth for computer terminal equipment, (approximately 40-50% per year). The writer knows of no other market available at the present time which offers this rate of growth.

The major computer manufacturers have in recent years placed their emphasis on larger and larger high speed computers. This has resulted in costs which are prohibitive for the typical small business. The high

---

*Report by Mr. Glenn Andrews and Mr. Fitzroy Kennedy, Arthur D. Little Company.

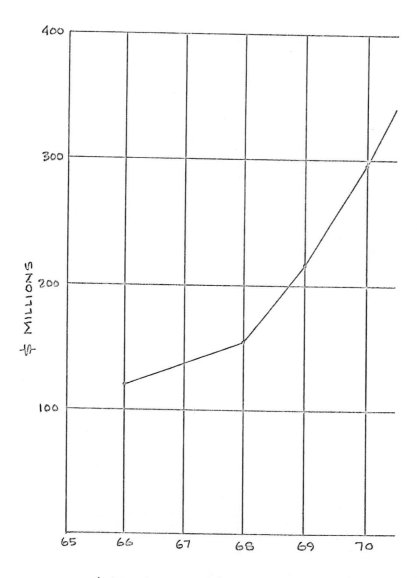

ANNUAL MARKET — TERMINAL & CONTROL EQUIP.
FIG. 1

FROM ADL REPORT

cost of these greatly sophisticated machines has dictated the use of "Time Sharing" in order for a user to utilize the machine's capability. Thus the real bottle-neck in controlling the information explosion is the application of suitable terminal equipment to gain entry to these computers, and to obtaining useful data from them.

Thus we are confident that a company which has a foundation of outstanding technical capability can take maximum advantage of the market which is now developing.

III. Personnel

An investment in a technologically oriented company is worth little more than the capabilities of its technical personnel. In Computer Terminal Corporation three of the best technically qualified people in their field will form the nucleous of a team dedicated not only to technical excellence but to practical realizable economic growth. Mr. Skelton was one of the pioneers in the application of the transistor to practical circuitry and later directed the design of the first digital computer utilizing integrated circuits.

Mr. Ray has personally designed data handling equipment in use on almost every U.S. Missile in the free world's arsenal and in addition he is currently directing work in digital communications for a major classified government program. He has been an active and prolific inventor as evidenced by the number of original designs which he has created.

Mr. Roche is a foremost authority in the field of digital communication systems. Mr. Roche developed the PCM Bit Synchronizer for use in the Atlantic Missile Range and is currently a specialist in the area of digital network synchronization and digital switching concepts.

Detailed resumes of each of these key personnel are included in this section.

Charles W. Skelton, President, Chief Executive Officer and Marketing Manager

Education:

Mr. Skelton received his BS degree in Electrical Engineering from Texas A&M University in 1950 and has done graduate work in Electrical Engineering at Southern Methodist University and in Psychology at the University of Texas at Arlington. His service education includes the Navy Radio Materiel School in 1942 and Navy Loran School in 1944.

He attended the American Management Association course in General Management in New York City in 1958, the AMA's Course for Company Presidents in 1964.

Professional Experience:

Mr. Skelton joined Communication Engineering Company in 1950 and worked with the development of VHF antennas of the parasitic and driven-array varieties. In 1952, he joined the engineering staff of Texas Instruments and was assigned to the development of a VHF receiver-transmitter for seismograph use. As a project engineer in 1954, he guided the development of a radio data receptor for the Navy. He was then successively project engineer for the development of an Air Force world-standard timing system, a digital intervalometer for the Air Force, a small computer for the Navy, a flight programmer for the Titan ICBM, and a nose-cone re-entry programmer for the Titan missile. Then, as Head of the Digital Circuits Section, he was responsible for the production phases of those projects. He directed the development

- 2 -

and production of flight programmers for the Titan II missile, the development of the flight programmers for the Pershing missile, a group of classified systems for the Air Force Technical Intelligence Center, and classified intelligence acquisition equipment for the Central Intelligence Agency. In 1959, as branch head in the missile department, he directed programs for the "worst case" circuit analysis for the Minuteman guidance computer.

Mr. Skelton resigned from Texas Instruments in 1960 to found International Data Systems, Inc. As President, he directed the marketing, design, and manufacture of airborne digital data handling systems such as multiplexing equipment for Saturn I and Saturn IB; low level signal switching equipment for the RS-70; signal conditioning equipment for Saturn, Scout, Minuteman, and other programs; complete pulse code modulation systems for project Sleigh-Ride, the Lance missile, the Hibex booster; and a high speed digital data acquisition system for Titan III.

Mr. Skelton joined Geotech in 1966 as a Senior Engineer in the Automatic Controls Branch; he is currently Manager of the Short Range Detection Department, responsible for marketing and engineering of classified equipment for the Department of Defense. He currently holds an active "Top Secret" clearance with the Department of Defense.

Honors and Patents:

In 1959, Mr. Skelton was the recipient of the faculty designated outstanding ex-student award of Navarro College. He is the grantee of 7 patents, primarily in the digital computer field, and is a Registered

Professional Engineer in the State of Texas, and a Senior Member of the IEEE.

References:

Gene Bishop, Senior Vice President, First National Bank in Dallas, President, First Dallas Capital Corp., Dallas, Texas
AC 214 RI9-4495

Melvin Goldstein, President, Technical Associates, Inc., 4521 W. Napoleon, New Orleans, La.
AC 504 888-4884

Elmer W. Dobbins, Investment Banker, Rotan Mosle, San Antonio, Texas
AC 512 CA3-3051

J. P. Ray, Vice President and Operations Manager

Education:

Mr. Ray received his BSEE degree in Electrical Engineering from the University of Texas in 1957. He took special courses in Digital Computer Engineering at the University of Michigan in 1959.

Professional Experience:

In 1957 he joined Texas Instuments, Inc. where he managed projects involving missile telemetry and flight programmers. He was credited with several important contributions to the fields of electronic speech analysis and speech coding. In 1960 he joined International Data Systems, Inc. where he was Assistant Vice President, Manager of Engineering and responsible for the Engineering Department primarily in the development of telemetry products for Aerospace applications. His major projects included Lance, Sleigh-Ride, Hibex and Saturn PCM systems.

In 1965 he joined Dynatronics as a Senior Staff Engineer concerned with directing the company's technical effort primarily in the area of advanced aerospace telemetry. He is presently Technical Program Manager and responsible for the design and development of advanced PCM Telemetry and communication equipment for the Model 35 Spacecraft.

Honors and Patents:

    Patent granted: "A Unique Binary Memory."
    Patents pending: "A Wide Range Voltage Regulator"
                   "A Magnetic Automatic Gain Control"
                   "Analog to Digital Converter"
                   "A High Speed Digital Alpha-Numeric Display System"

- 2 -

"A Digital Frequency Synthesizer".

He is a member of American Management Association.

References:

H. M. Merridith,	Vice President, First National Bank in Dallas, Dallas, Texas AC 214 RI9-4011
Darrell Lafitte,	President, North Dallas Bank and Trust, 11811 Preston, Dallas, Texas AC 214 EM8-2856
Joe Burch Gilbert,	Gerald Hines Co., 2114 Pelham, Houston, Texas AC 713 JA3-8663
W. F. Donnell	Senior Engineer, Tracor, Inc., Austin, Texas

Austin O. Roche, Vice President and Technical Director

Education:

Mr. Roche received his BSEE at Purdue University with graduate studies at the University of Florida.

Professional Experience:

Mr. Roche joined Radiation, Inc. as an Engineer and participated in the design and development of a PPM/AM telemetry system, and and S-band signal generator and test set. He was responsible for development of various fluid process controls as an Engineer with Hoffman Speciality Manufacturing Corporation. He joined Emerson Research Laboratory as an Engineer and participated in the design of airborne pseudo-doppler altimetry systems (including the AN/APM-100 system), radio frequency proximity fusing, transistorized digital control circuitry for automatic postal apparatus, pulse pattern generators for high speed telegraphy testing, and summing magnetic amplifiers for airborne computers.

He joined Dynatronics as a Group Engineer in the Data Systems Section responsible for the design and high capacity switching regulated power supplies and associated load transfer equipment for AMR. As a Project Engineer he was responsible for concept and development project for USAF-CRC high altitude research balloon encoders multiplexers and transmitters. This project entailed extremely low level signal processing techniques. He was then assigned as Project engineer for initial development of the PCM bit synchronizer for AMR (TLM-62) and for the AMR ballistic camera synchronization and timing system.

As a Section Head in the Special Development Projects he was responsible for development of telemetry pre-detection recording systems for AMR and ARIS ships. This program included development of modular switchable bandwidth receivers, pre-detection converters and special signal combiners. Mr. Roche was responsible for the development of standard product PCM bit synchronization equipments, range safety officers consoles, chopper stabilized amplifiers, and active filter design for signal conditioning systems.

As a Senior Staff Engineer/Manager Research Staff at Dynatronics he was responsible for and participated in contract and company studies in Deep Space Telemetry and UHF Telemetry Tracking and Acquisition studies, company sponsored studies in advanced synchronization technique, PFM synchronization and demodulation. Mr. Roche also developed systems for acquisition and telemetering of wideband data. Major participation in numerous proposal activities were: stored program decommutation, advanced tracking and acquisition systems, ultra-linear receiver systems, signal design and modulation aspects, information processing and data compaction systems, signal conditioning and pre-sampling data filters. He was responsible for conception and development of DSB-AM/FM wideband telemetry system and the design and development of Demodulators for Lunar Orbiter Program.

In the Advanced Communications Department in Martin-Marietta Corp. as a Senior Staff Engineer he served in analysis in the area of digital communication network synchronization and digital computer

- 3 -

switching concepts, and performed analysis in time and frequency division multiple access techniques. He participated in pre-proposal study of advanced army tactical communication system. He held major proposal responsibility in multiple access tactical satellite and tactical VHF radio modulation studies.

Mr. Roche joined Dynatronics again and was responsible for directing company's technical effort primarily in the area of advanced aerospace telemetry and communication techniques. He is a Senior Staff Engineer and is responsible to Engineering Vice President for the coordination of staff activities in the engineering department.

Publication and Affiliations:

"The Use of Double Sideband Suppressed Carrier Modulation as a Subcarrier for Vibration Telemetry" National Telemetering Conference, 1964.

"Applications of DSB/FM to Vibration Telemetry" Proceedings, International Telemetering Conference, 1965.

He is a member of American Ordnance Association and the IEEE.

References:

Neil D. Skinner, President, Hoffman Manufacturing Co., 1700 W. 10th St., Indianapolis, Indiana.
AC 317 ME2-7546

B. S. Chen, Program Manger, Pan-American World Airways, Cocoa Beach, Florida.
AC 305 494-4122

Charles J. Barnhill, Lederer, Barnhill, and Fox, First National Bank Bldg., 33 South Clark, Chicago, Illinois
AC 312 236-1224

- 4 -

Victor D. Poor, Vice President, Fredrick Electronics Corp., P.O. Box 502,
Fredrick, Maryland
AC 301 662-5901

IV. Financial Projections

Included in this section are Table 1, a Pro-Forma Operating Statement for the first year's operation and Table 2, Pro-Forma Balance Sheets for the first five years. Full provision is made for corporate income tax in the projections, and it is anticipated that all initial sub-ordinated convertible notes would be converted at the end of the third year.

A break-down of paid in capitalization including both common stock and sub-ordinated convertible debt is shown in Table 3. It is proposed that the convertible notes be sub-ordinated in favor of bank debt only, thus they would be senior to common stock. It is proposed that the notes bear an interest rate of 6%, and be convertible into common stock on the basis of $35.00 per share at the option of either the noteholder or the company. The notes would mature in five years.

PRO FORMA OPERATING STATEMENT 12 MONTHS

Sales		500,000
Cost of Sales		
Material	190,000	
Direct Labor (Salaries and Wages)	71,000	
Overhead:		
Engineering Salaries and Wages	21,000	
Payroll Taxes	5,000	
Insurance	3,000	
Equipment Rental	7,000	
Freight	1,500	
Utilities	1,800	
Miscellaneous Repairs and Supplies	2,000	
	302,300	
Gross Operating Profit		197,700
General and Administrative Expense		
Sales Commission	50,000	
Marketing Manager's Commission	8,000	
Salaries	29,600	
Other Expenses		
Advertising	5,000	
Office Equipment	1,000	
Utilities	600	
Insurance	500	
Professional Services	3,000	
Rent	3,600	
Supplies	1,500	
Taxes and Licenses	1,000	
Telephone and Telegraph	4,000	
Travel and Entertainment (40 Man-Trips)	10,000	
	117,800	
Net Operating Profit		79,800
Provision for Federal Income Tax		41,800*
Net Operating Profit after Taxes		38,000
Earnings per Share		1.83
Estimated Offering Price per Share based on price to earnings ratio of 20/1		37.00

*Based on restoration of 52% Federal Corporate Tax, under current tax rate, income tax would be 31,256

TABLE I

PRO FORMA BALANCE SHEET 5 YEARS

Sales	500	1000	1900	2600	3300
Assets:					
Cash	351	280	572	734	930
Accounts Receivable	135	248	531	725	920
Inventory	180	250	475	591	750
Other	10	15	20	20	20
Total Current Assets	496	793	1598	2070	2620
Property Net	37	62	98	140	199
Research & Development	59	75	92	105	100
Total Assets	592	930	1788	2315	2999
Liabilities:					
Accounts Payable & Accrued	154	280	533	730	925
Equipment Notes	-	20	100	110	120
Bank Notes	-	-	350	430	613
Total Current Liability	154	300	983	1270	1658
Long Term Liabilities	350	350	350*	-	-
Stock Holder Equity	150	150	150	500	500
Retained Earnings	38	130	305	545	841
Total Liability & Equity	592	930	1788	2315	2999
Current Ratio	3.44	3.1	1.6	1.63	1.64
Earnings per share	1.83	4.50	8.00	8.00	9.60
Estimated Value per share based on P/E of 20/1	37.00	90.00	160.00	160.00	192.00

*Convertible debt would be converted to equity during 4th year, assuming conversion price of 35.00 per share.

TABLE 2

BREAK-DOWN OF PAID IN CAPITALIZATION, INCLUDING SUB-ORDINATED NOTES

Shares offered to initial investors @ $10.00/share	15,000
Shares offered to promoters in consideration of equipment design	4,200
Underwriters commission payable in stock	1,500
Total shares outstanding	20,700
Total paid in equity capital	$ 150,000
Convertible sub-ordinated 6% notes convertible at 35.00 per share	350,000
Total equity and debt capitalization	$ 500,000

TABLE 3

POSITION	QUANTITY		ANNUAL PAYROLL	DIRECT		OVERHEAD		G & A	
	6 MO.	1 YR.	$/YR.	6 MO.	1 YR.	6 MO.	1 YR.	6 MO.	1 YR.
President	1	1	15,000*	-	-	-	-	7,500	15,000
Vice Pres., Oper. Mgr.	1	1	21,000	4,000	8,000	6,500	13,000	-	-
Vice Pres., Tech. Dir.	1	1	20,000	3,500	7,000	6,500	13,000	-	-
Manufacturing Mgr.	1	1	16,000	-	6,000	-	2,000	-	-**
Controller	1	1	11,000	-	-	-	-	5,500	11,000
Engineer	2	2	20,000	8,500	17,000	1,000	3,000	-	-
Technicians & Draftsmen	4	4	26,000	12,000	24,000	1,000	2,000	-	-
Clerk Typist	2	2	7,600	1,000	2,000	1,000	2,000	1,800	3,600
Shipping & Receiving	-	1	4,000	-	-	-	4,000	-	-
Test Technician	-	1	5,000	-	5,000	-	-	-	-
Assembly	-	10	36,000	-	36,000	-	-	-	-
			181,600	29,000	105,000	16,000	39,000	14,800	29,600

*In addition to salary for President shown, he would also receive 10% of net profit before taxes.

**Last six months only.

Distribution of salaries and wages 6 months and 1 year.

TABLE 4

CASH RECEIPTS AND DISBURSEMENTS

Months	1	2	3	4	5	6	7	8	9	10	11	12
Starting Cash	500	482.6	461.5	442.4	416.0	379.5	339.9	295	255	225.3	234.4	279.7
Material	3	7	5	10	20	20	30	30	30	30	35	40
Salaries & Wages	11	11	11	11	13	16	16	16	16	16	16	16
Payroll Taxes	.4	.4	.4	.4	.5	.6	.6	.6	.6	.6	.6	.6
Insurance	.2	.2	.2	.2	.3	.3	.3	.4	.4	.4	.4	.4
Equipment Rental	.6	.6	.6	.6	.7	.7	.7	.7	.7	.7	.7	.7
Freight	.12	.12	.12	.12	.12	.12	.12	.12	.12	.12	.12	.12
Utilities	.2	.2	.2	.2	.2	.2	.2	.2	.2	.2	.2	.2
Supplies & Repairs	.3	.3	.3	.3	.3	.3	.3	.3	.3	.3	.3	.3
Advertising				2.0			1.0			1.0		1.0
Professional Services	.25	.25	.24	.25	.25	.25	.25	.25	.25	.25	.25	.25
Taxes & Licenses	.25			.25			.25			.25		
Telephone & Telegraph	.3	.3	.3	.3	.3	.3	.3	.3	.3	.3	.3	.3
Travel & Entertainment	.8	.8	.8	.8	.8	.8	.8	.8	.8	.8	.8	.8
President's Commission												
Total Disbursements	17.4	21.1	19.1	26.4	36.5	39.6	50.9	49.7	49.7	50.9	54.7	68.7
Ending Cash Less Revenues	482.6	461.5	442.4	416.0	379.5	339.9	290	245	205.3	174.4	179.7	211
Revenues	-	-	-	-	-	-	5	10	20	60	100	140
Ending Cash	-	-	-	-	-	-	294.3	255	225.3	234.4	279.7	351

TABLE 5

V. Conclusion

Because of the very nature of the computer terminal market, it ap appears highly doubtful that the venture could be financed for the long term by private investors. The writer, therefore, anticipates that a public offering of stock would be made, hopefully after the first year's operation, and certainly when a satisfactory profit picture is established. The current demand for common stock in the computer oriented companies is high, with most such stocks selling at very high price to earnings multiples. However, the projections contained in this proposal with respect to anticipated value per share are based on a conservative 20/1 price to earnings multiple. Thus the potential capital gain for an initial investor could far exceed the trend shown in Table 2.

Bibliography

ARCNET Trade Association, general information at http://www.arcnet.com/abtarc.htm, accessed November 5, 2009.

Author interviews or personal communication with (in random order) Dave Gust, John Murphy, Jack Frassanito, Vic Poor, Harry Pyle, Jonathan Schmidt, Stan Mazor, Hal Feeney, Federico Faggin, Richard Erickson, Michael Fischer, Gerald Mazur, Gerry Cullen, Ed Gistaro, Herb Baskin, Asher Edelman (in 1986), Andy Viger, Austin Roche, Chris Roche, Bob McDowell, Joel Norvell, John Bender, Bob McClure, Mike Green, Gordon Peterson, Ted Hoff, Ted Nelson, David Monroe, Robert Metcalfe, Chuck Miller, Paul Ceruzzi, Amy Wohl, and Chrisa Norman Scoggins.

Bureau of Economic Analysis, U.S. Department of Commerce, www.bea.gov, various tables, accessed March 2, 2009.

Businessweek, 12-10-79 and 9-23-85.

Campbell-Kelly, Martin, and Aspray, William, "Computer: a history of the information machine," Basic Books, 1996.

Ceruzzi, Paul E., "A History of Modern Computing" MIT Press, 2003.

Christensen, Clayton M., "The Innovator's Dilemma: when new technologies cause great firms to fail," Harvard Business School Press, Boston, 1997.

Communications of the ACM, http://portal.acm.org/citation.cfm?id=360253&coll=ACM&dl=ACM&CFID=50907435&CFTOKEN=49315255, accessed November 5, 2009.

Computer History Museum, interview with Hal Feeney, June 6, 2004.

Computer History Museum, interview with Harry Pyle, May 27, 2005.

Computer History Museum, interview with Vic Poor, December 8, 2004.

Computer History Museum, interviews with Federico Faggin, 2004-2005.

Computer History Museum, September 21, 2006, group interview with Federico Faggin, Hal Feeney, Ed Gelbach, Ted Hoff, Stan Mazor, and Hank Smith.

Computer History Museum, http://www.computerhistory.org/semiconductor/timeline/1968-SGT.html, assessed June 30, 2009.

Computer Industry Almanac Inc. and Egil Juliussen

Computer Terminal Corp., annual reports.

Computerworld, stories that appeared on 4-22-85, 3-31-86, and 3-9-87.

Cullen, Gerry, "The Coldest Call," 2007, Austin, Texas.

Datamation, January 1976, and 5-1-85.

Datapoint Corp., "Datapoint World" (marketing newsletter), July 1983.

Datapoint Corp., "vanity" corporate history written in 1981 and abandoned in 1982, author unknown.

Datapoint Corp., various annual reports, product catalogs and press releases, and brochures.

Dun & Bradstreet, report on Computer Terminal Corporation, dated 1-21-70.

Electronics Weekly, history series, summer 2008.

Electronics, 11-23-70 and 6-7-71.

Feeney archives.

Fischer archives.

Forbes Magazine, 12-10-79 and 12-2-85.

Fortune Magazine, November 1985, and (for Fortune 500 positions) Fortune web site archives.

Frassanito archives.

Freiberger, Paul, and Swaine, Michael, "Fire in the Valley," McGraw-Hill, 1984.

Gartner Inc., http://www.gartner.com/it/page.jsp?id=703807, accessed 3-11-2009.

General Mills, http://www.generalmills.com/corporate/company/hist_SearchableBook.pdf, accessed August 7, 2009.

Gust archives.

IEEE, Computer Group News, July-August 1970

IEEE Solid-State Circuits Magazine, Winter 2009.

Intel Corp., "A Revolution in Progress," 1984 (16-year anniversary brochure).

Intel Corp., http://www.intel.com/museum/archives/4004.htm, accessed July 22, 2009.

Intellectual Property Owners Association, http://www.ipo.org/AM/Template.cfm?Section=19981&Template=/CM/ContentDisplay.cfm&ContentID=3674, and http://www.ipo.org/AM/Template.cfm?Section=Search&template=/CM/HTMLDisplay.cfm&ContentID=10692, accessed September 10, 2009.

Intel Corp., 2008 Annual Report.

MIS Week, 3-14-84 and 1-16-85.

National Weather Service, http://www.srh.noaa.gov/ewx/html/wxevent/Climate_Narratives/julclimate.htm, referenced April 28, 2009.

Network World, 12-4-89.

New York Times, 2-19-82, 5-21-91, 11-7-91, and 10-16-2002.

Roche family archives.

San Antonio Express-News archive undifferentiated clippings (from either the Express-News or the Light, as the archives merged in 1992 when the Hearst Corp. acquired the Express-News and closed the Light) stories from 3-4-73, 11-25-73, 6-9-74, 2-13-75, 2-18-75, 3-13-77, 5-6-80, 6-11-80, 1-5-81, 6-26-81, and 2-7-83.

San Antonio Express-News, stories appearing on 7-11-71, 12-24-72, 8-5-73, 3-2-75, 2-5-76, 11-13-77, 7-24-78, 11-28-79, 11-21-80, 9-10-81, 2-12-82, 2-19-82, 4-8-82, 4-9-82, 4-25-82, 6-18-82, 2-5-85, 3-30-85, 5-3-85, 6-14-85, 7-2-85, 7-31-85, 9-6-85, 9-7-85, 9-10-85, 4-12-86, 6-17-86, 10-21-86, 2-28-87, 9-4-87, 10-13-87, 12-13-87, 4-12-89, 4-17-90, 12-20-90, 1-4-91, 4-12-91, 7-20-91, 10-9-92, 8-3-94, 10-5-94, 11-15-94, 11-24-94, and 4-16-95.

San Antonio Light, multi-part Datapoint history series by Richard Erickson, September 1986.

San Antonio Light, stories appearing on 8-6-78, 9-2-79, 10-13-81, 2-4-82, 5-22-82, 4-3-85, 12-18-86, 2-2-87, 3-3-87, 3-8-87, 4-21-88, 12-9-89, 3-11-90, 7-10-90, and 10-11-91.

San Antonio Magazine (San Antonio Chamber of Commerce), August 1974.

San Antonio Public Library, vertical files concerning Datapoint and Intelogic Trace.

Schaller, Robert, dissertation titled "Technological Innovation in the Semiconductor Industry," George Mason University, 2004.

Seybold Report on Office Systems, April 1982, volume 5 number 4.

Shiller, Robert J., "Irrational Exuberance," second edition, Doubleday Business, 2006, plus P/E figures from the book posted at www.econ.yale.edu/~shiller/data/ie_data.xls.

Southwest Technical Products Corp., http://www.swtpc.com/mholley/History/SWTPC_History.htm, accessed March 25, 2009.

Stanford University, Ted Hoff interview, transcribed at http://www-sul.stanford.edu/depts/hasrg/histsci/silicongenesis/hoff-ntb.html.

Stott, Martha, "Two Decades of Networking," ARCNET Works newsletter, ARCNET Trade Association, Fall 1998.

Time Magazine, 12-11-89.

U.S. Patent and Trademark Office web site.

Usenet FAQs, http://www.faqs.org/faqs/dec-faq/pdp8-models/section-2.html, accessed July 3, 2009.

Veeneman, Dan, http://hp9100.com/, accessed July 10, 2009.

Wall Street Journal, stories on 7-23-69, 5-27-70, 12-14-71, 2-29-72, 8-15-72, 9-28-72, 12-8-72, 3-15-73, 6-1-73, 8-2-73, 7-17-74, 7-14-75, 2-27-76, 4-22-76, 4-4-77, 4-18-77, 12-15-77, 3-23-78, 9-10-80, 9-23-80, 9-29-80, 10-24-80, 2-17-81, 4-2-81, 5-29-81, 6-1-81, 11-4-81, 2-4-82, 2-19-82, 3-24-82, 5-3-82, 5-13-82, 5-14-82, 5-27-82, 10-29-82, 11-19-82, 2-15-83, 5-16-83, 6-5-84, 6-19-84, 7-27-84, 9-27-84, 12-10-84, 12-11-84, 1-11-85, 1-14-85, 1-25-85, 1-30-85, 3-11-85, 3-12-85, 3-18-85, 4-1-85, 6-14-85, 9-5-85, 1-16-87, 3-25-87, 6-2-87, and 9-4-87.

Welsh, David and Theresa, "Priming the Pump," 2007, The Seeker Books, Ferndale, MI.

Wikipedia, "TRS-80," accessed December 14, 2009.

Wood, Lamont, list of Datapoint press releases compiled on July 29, 1986, of the press releases then on file in the Datapoint product publicity department, for a magazine article. If the department had run out of copies, the release was not on file. (Only a few of the actual press releases survive at this writing.)

Wood, Lamont, notes on Peat Marwick Mitchell v. Datapoint, made in 1986.

Datapoint 2200 Product system introduction, J.P. Ray, V. D. Poor, J. Everett, A. O. Roche

Freedom line printer

Datapoint 2200 first production circa 1970, photo: James Whitcomb

Datapoint 5500 office automation suite, photo: James Whitcomb

George Lois add campaign circa 1970

Datapoint product concept circa 1981, photo: John Dyer

Intel/Datapoint 8008 microprocessor

United States Patent Office Des. 224,415
Patented July 25, 1972

224,415
COMPUTER TERMINAL
Jon P. Ray and Austin O. Roche, San Antonio, Tex., and John R. Frassanito, Oyster Bay, N.Y., assignors to Computer Terminal Corporation, San Antonio, Tex.
Filed Nov. 27, 1970, Ser. No. 26,176
Term of patent 14 years
Int. Cl. D14—02
U.S. Cl. D26—5

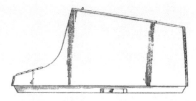

FIG. 1

FIG. 2

THIS IS TO CERTIFY that this is a true copy from the records of the U. S. Patent and Trademark Office of the first page of the above identified patent:

Certifying Officer   1/18/08 Date

Datapoint 2200 patent

J. Phillip Ray, Gerald Mazur, Austin O. Roche

R.O. Norman, R. Fogg

Datapoint 2200 illustration used in financial presentation that raised the development funding, John Frassanito

Datapoint 2200 illustration of word processing system, John Frassanito

Laser printer optics bench

Laser printer design mockup

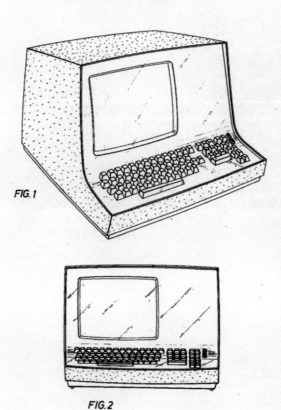

Datapoint 3300 Patent

Datapoint 3300 Concept illustration circa 1968

Datapoint 5500 system

Modular workstation design

JF&A1, one of five prototype videoconferencing systems developed by John Frassanito & Associates, David Monroe U.S. Patent number 4,710,917, 4,847,829.

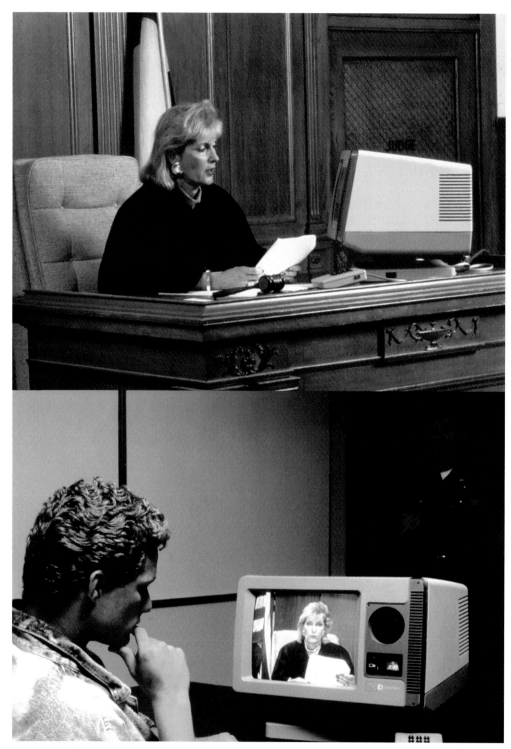
Production Datapoint MINX system installation

Conceptual Avatar Teleconferencing Robot

Production Datapoint LightLink Infrared data transmission system installation

Jack Frassanito, Flo, Vic Poor and in Mexico

Jack, Phil, Gus, at McClure meeting

Jack Frassanito Mike Faherty

David Monroe Harold E. O'Kelly circa 1978

Jon Phillip Ray

Phil Ray, Bob Bauer, Gus Roche, Mnemonics founders meeting

Dr. Carver Meade, Jack Frassanito

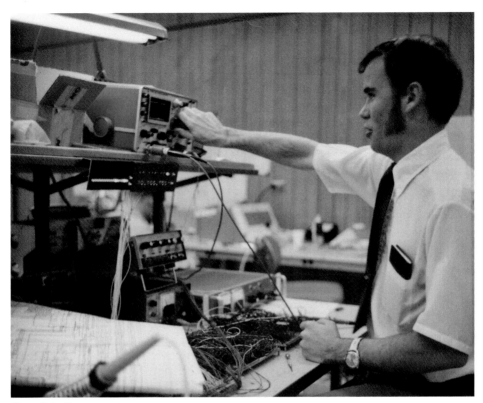
Gary Asbell

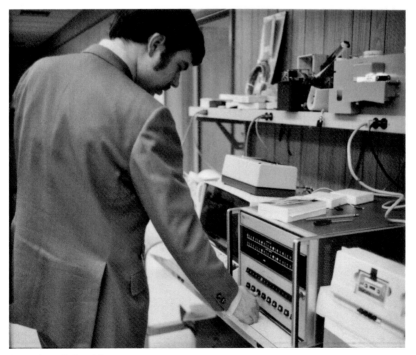
Jonathan Schmidt

Harry Pyle

Dr. Ted Hoff

Gus Roach, Dr. Bob McClure,

David Monroe, Jonathan Schmidt

Gordon Peterson   Dr. Robert Noyce

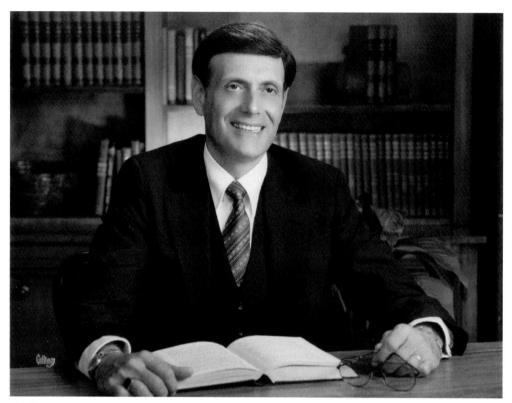
Ed Gestaro

Dr. Federico Faggin

9725 Datapoint Drive, Corporate Headquarters circa 1970

A. O. Roche Dedication plaque